2014

HTML & CSS - CURVE

HTML or Hyper Text Markup Language is used to create web pages. HTML elements form the building blocks of all web pages. Learn the basics of HTML & CSS and create your own web pages in 3 days.

By SANTOSH C J

HTML & CSS - Curve

Acknowledgments

This book contains the work of many individuals who gave technology and its children a different name and face over a period of time. I wish this book helps many aspirants who would like to learn HTML & CSS in a short period of time.

This book is dedicated to my parents C Jayaraman and D. Uma Devi. I would like to thank my brothers Magesh C J and Satish C J who helped me in making my dream a reality.

Special thanks to all my friends across the globe.

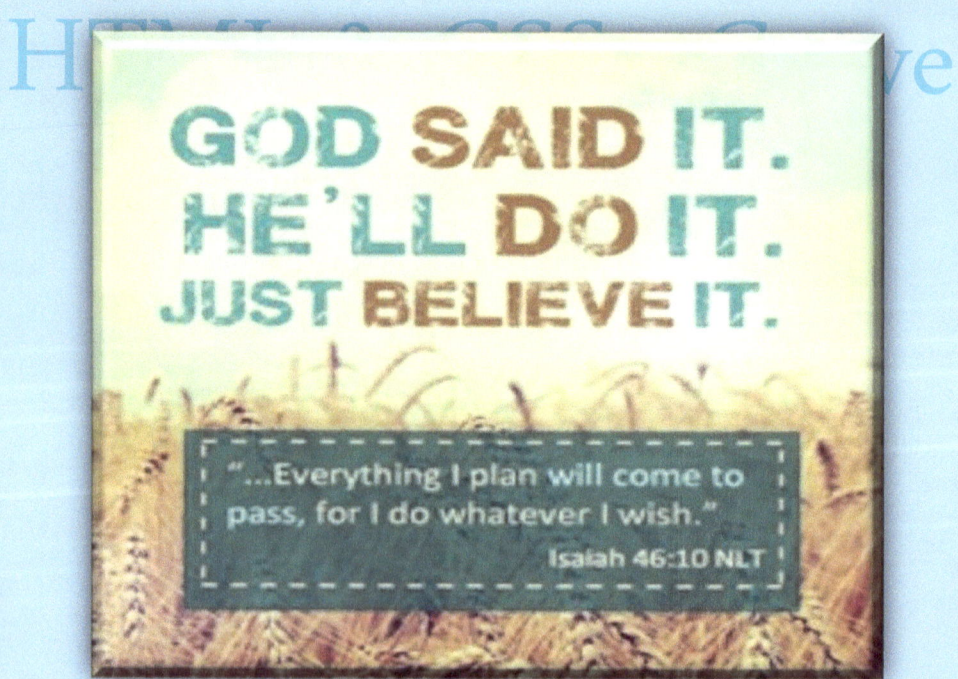

INDEX

HTML & CSS - Curve

INTRODUCTION

HTML or Hyper Text Markup language is a Markup language used to create web pages. HTML elements form the building blocks of all web pages. An HTML file is a text file containing small markup tags. These tags tell the browser how to display a web page. Browsers like Internet Explorer, Mozilla Firefox, and Google Chrome are used to read HTML files. These browsers hide the HTML tags and display only the content of the page. An HTML file must have an .htm or .html extension.

In order to create a HTML file, the user has to either download HTML editors like HTML-Kit-Tools, Robohelp html or the user can use NOTEPAD, a common text editor application that is available in all versions of windows. The user can also use HTML online edition to create/edit html files. In case, if the user is using Notepad to create a HTML file, the user has to save the file with a .htm or .html extension. The .htm and .html extensions have few differences. Initially few types of software that was used to create html file has limitations and allowed the user to save the created file as a .htm extension. Both .htm and .html extension are treated as different files by the browsers.

Best way to learn HTML can be achieved by understanding how other people have coded their html pages. This can be done by visiting any of the website, doing a right click and selecting the option "view page source". The page source will be shown in a new web page.

CSS - Cascading Style Sheets is a style sheet used in HTML language. It is used for creating and formatting the content of a document written in a markup language.

Basic Syntax:

CSS has a simple syntax and uses a number of English keywords to specify the names of various style properties.

A style sheet consists of a list of rules. Each rule or rule-set consists of one or more selectors, and a declaration block

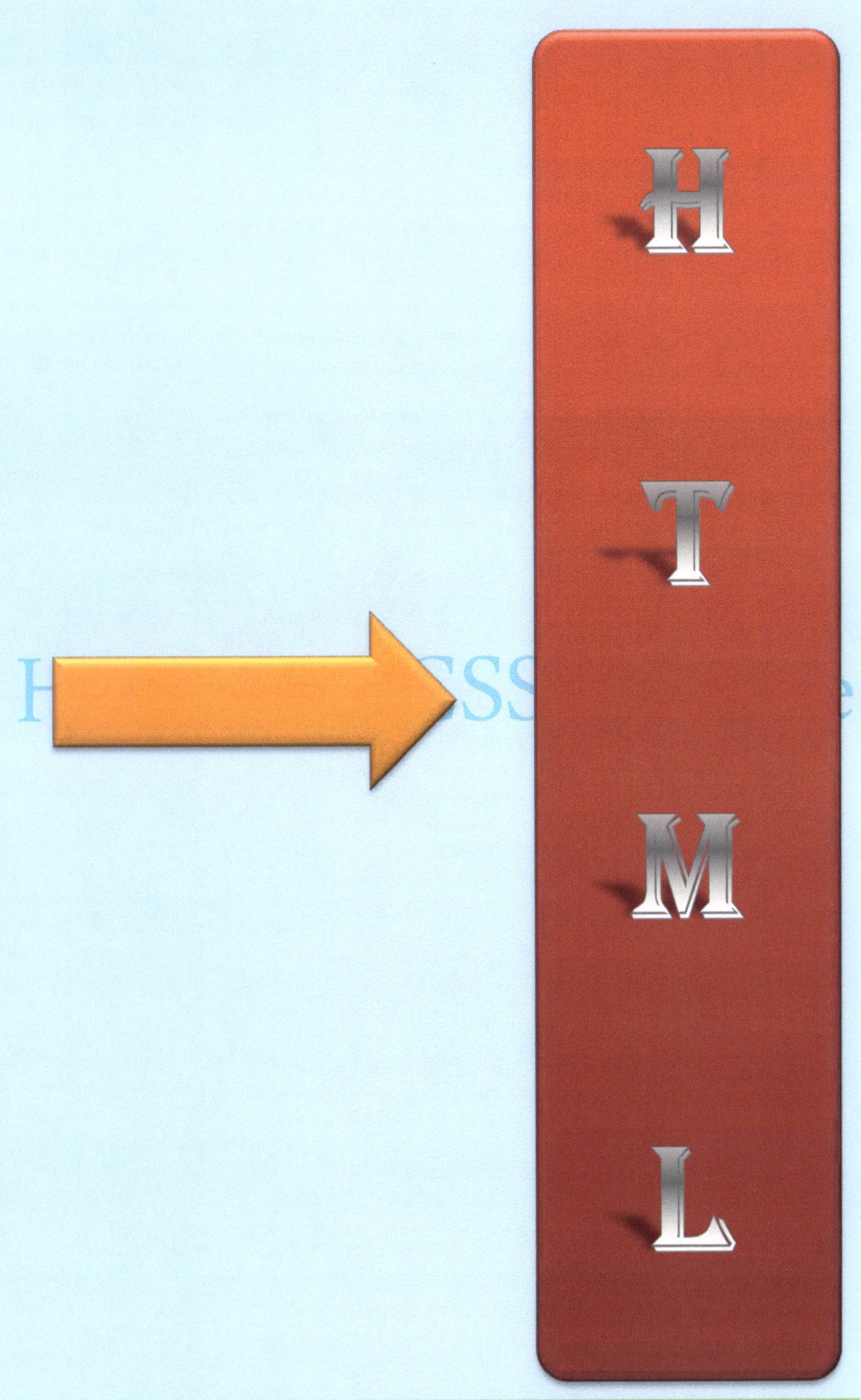

HTML

Hypertext Markup Language - Used to design Webpages.

HTML Tag rules:

- HTML tags are used to mark-up HTML elements
- HTML tags are surrounded by the two characters < and >
- The surrounding characters are called angle brackets
- HTML tags normally come in pairs like <b> and </b>
- The first tag in a pair is the start tag, the second tag is the end tag
- The text between the start and end tags is the element content.
- HTML tags are not case sensitive, <b> means the same as <B>

Important Components of a HTML file:

- <!doctype html>
- <html>
- <head>
- <title>
- <body>

<!doctype html>

It is the first declaration in the document. It is an instruction set to the web browser about the version of HTML used.

<html>

It is the root of the HTML document. It tells the browser that it an HTML document. It is the main container of HTML pages.

\<head\>

It is the container for page header information. It includes the title of the document. It can include meta information, styles and scripts.

\<title\>

It is the title of the page. One cannot have more than one title element in the HTML document.

\<body\>

It is the main body of the page. It contains all the contents of Text, Images, tables, links etc.

How to create a simple HTML documents using NOTEPAD:

- Open Notepad
- Type the contents
- Choose option file>save as> Enter the file name>Save as type – All files.
- Name the file with a .html extension like Webpage.html, doc1.html etc.
- Save

How to open an HTML file:

- Find the created HTML file with a .html extension
- Right click and choose browser to open the file. Like Internet explorer, Firefox, chrome etc.

Example: Simple html Document-1.01: Notepad:

```
<!doctype html>
<html>
<head>
<title>Training page</title>
</head>

<body>
<h1>Hello</h1>
</body>
</html>
```

HTML & CSS - Curve

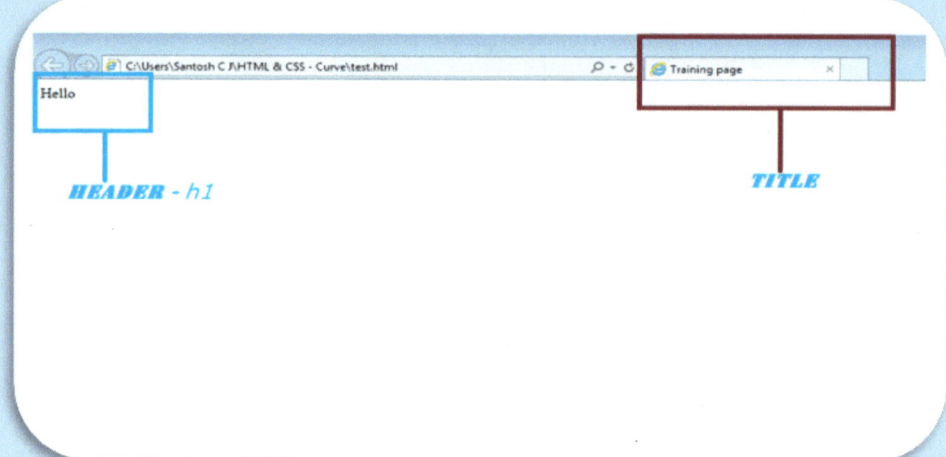

HTML Tags:

HTML language is a markup language and we use many tags to markup text. In HTML there are two types of tags 1) Logical tags and 2) Physical tags.

Logical Tags:

Logical tags give meaning to the closed texts. An example of a logical tag is the <strong> </strong> tag. By placing text in between these tags you are telling the browser that the text has some greater importance.

Physical Tags:

Physical tags on the other hand provide specific instructions on how to display the text they enclose. Physical tags were invented to add style to HTML pages.

Examples of physical tags include:
- <b>: Makes the text bold.
- <big>: Makes the text usually one size bigger than what's around it.
- <i>: Makes text italic.

Nested Tags:

When you enclose an element in multiple tags, the last tag opened should be the first tag closed.

For example:
- <p><b><em> *Not the proper way to close nested tags*.</p></em></b>
- <p><b><em>P*roper way to close nested tags.* </em></b></p>

Further Html tags can be classified in to Basic Tags, Meta tags, Formatting Tags, Forms and input tags, Frames, Images, Audio & video, Links, Lists and Tables.

Basic Tags:

Basic tags are tags that form the basic structure of a HTML document.

BASIC TAGS	
TAG	**DESCRIPTION**
<!DOCTYPE>	Document type
<html>	HTML document
<title>	Title for the document
<body>	Document's body
<h1> to <h6>	HTML headings
<p>	Paragraph
 	Single line break
<hr>	Draws a straight line
<!--...-->	Comment

The desire of knowledge, like the thirst for riches, increases ever with the acquisition of it. – Laurence Sterne

EXAMPLE- BASIC TAGS: 1.02 – NOTEPAD:

```html
<!doctype html>
<html>
<head>
<title>Training page</title>
<meta name="Test" content="Basic Tags explanation"/>
</head>
<body>
<h1>Welcome to Training page<h1>
<p>This training page will explain how HTML basic tags are used.<br>
This is Sam.<br>
This is a training page.<br>
<hr>
To draw a straght line.
</p>
</body>
</html>
```

OUTPUT- 1.02 – INTERNET EXPLORER:

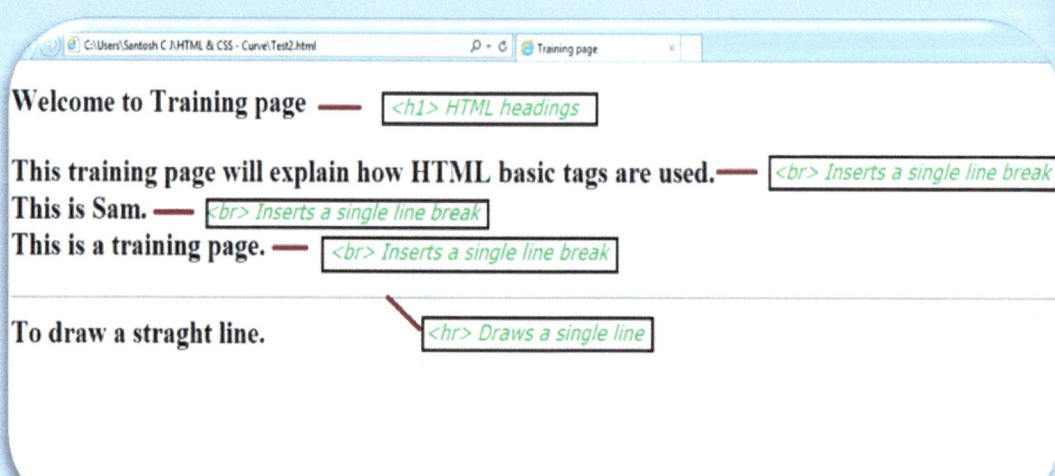

META TAGS:

HTML lets you specify metadata - information about a document rather than document content -in a variety of ways. The META element can be used to include name/value pairs describing properties of the HTML document, such as author, Expiry Date, a list of key words, author etc. The <meta> tag is an empty element and does not have a closing tag, a slash character has to be used at the end of the element.

META

TAG	DESCRIPTION
<head>	information about the document
<meta>	metadata about an HTML document
<base>	base URL/target for all relative URLs in a document

HTML & CSS - Curve

EXAMPLE: META TAGS-1.03: NOTEPAD:

```
<!DOCTYPE html>
<html>
<head>
<title>Training Page</title>
<meta name="Test" content="meta tag explanation" />
</head>
<body>
<h1>Hello</h1>
</body>
</html>
```

FORMATTING TAGS:

Formatting tags are used to format text content in a HTML document. The output of this can be seen in a web page. Some of the Formatting tags often used are:

FORMATTING	
TAG	**DESCRIPTION**
\<abbr\>	Abbreviation
\<address\>	Contact information for the author/owner of a document/article
\<b\>	Bold text
\<blockquote\>	A section that is quoted from another source
\<del\>	Deleted Text from a document
\<dfn\>	Definition term
\<em\>	Emphasized text
\<i\>	Italic
\<ins\>	)Inserted text
\<kbd\>	Keyboard input
\<mark\>	Highlighted text
\<pre\>	Preformatted text
\<progress\>	Task in progress
\<q\>	Short quotation
\<s\>	Text that is no longer correct
\<samp\>	Output from a computer program
\<small\>	Smaller text
\<strong\>	Important text
\<sub\>	Subscripted text
\<sup\>	Superscripted text
\<time\>	Date/time
\<u\>	Underlining text
\<var\>	Variable

EXAMPLE: FORMATTING TAGS-1.04: NOTEPAD:

```
<!doctype html>
<html>
<head>
<title>Training Page</title>
<meta name="Test" content="Formatting Tag explanation"/>
</head>

<body>
<h1>Welcome to Training Page - <u>FOMATTING TAGS</u></h1>

<p>
1) Abbreviation - <abbr title="World wide web">www</abbr> <br>
2) Contact information for the author/owner of a document/article - <address>
Santosh c J; Phone; 0000000000</address><br>
3) Bold text -<b> Hai</b><br>
4) A section that is quoted from another source - <blockquote> This is a training
module</blockquote> <br>
5) Deleted Text from a document - <del>This is Sam</del><br>
6) Definition term - Internet Explorer :<dfn>A browser that is used to see web pages.
</dfn><br>
7) Emphasized text - I love <em> the way you love me.</em> <br>
8) Italic - I hate to <i>lie</i><br>
9) Inserted text - Life is <ins>bigger </ins>than anything.<br>
10) Keyboard input - Enter your <kbd>Text here.</kbd> <br>
11) Highlighted text - Everybody says <mark> God is Great</mark><br>
12) Preformatted text - Smile <pre>when you have teeth.</pre> <br>
13) Task in progress - <progress>--</progress> <br>
14) Short quotation - Life is<q> TO LIVE</q><br>
15) Text that is no longer correct - Sometimes <s>i could</s> we would care for
you<br>
16) Output from a computer program - Output:<Samp> Be in Time</samp><br>
17) Smaller text - The proverb says <small>"Think twice before you
decide"</small><br>
18) Important text - The proverb says <strong> "Perseverance is the key to
victory"</strong><br>
19) Subscripted text - H<sub>2</sub>O<br>
20) Superscripted text - 10 <sup>2</sup>x11<sup>44</sup><br>
<time>21) Date/Time - <time> May;5:1983, 08:40 PM</time><br>
22) Underlining text - Welcome to <u> SAM STUDIOS</u><br>
23) Variable - <var> missing</var><br>
<hr>
</p>
</body>
</html>
```

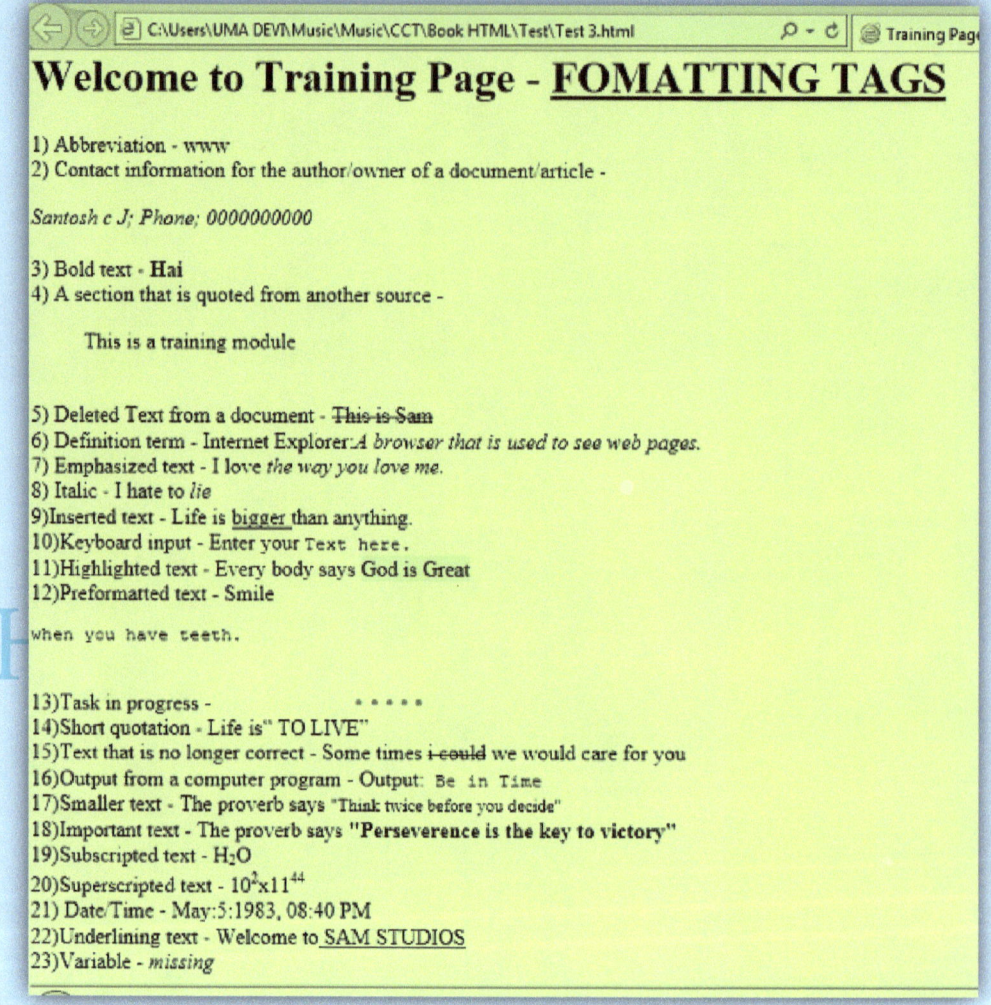

FORMS & INPUT Tags:

These tags are used to create forms and Input area so that a user can enter required info in a web page. A form can be divided in to three parts:

- Text input area
- Buttons
- Result area

Text Input area:

It is the input area where the User has to enter the data requested. It can be of different size and the character input can be limited.

Buttons:

Buttons are the important part of the form tags. There are different types of buttons like Submit, Search, query etc. The size of the buttons can be altered.

Result Area:

Some forms will have a result area where mathematical calculations are used. Some of the frequently used Forms and Input Tags:

FORMS & INPUT	
TAG	DESCRIPTION
<form>	HTML form for user input
<input>	Input control
<textarea>	Multiline input control or text area
<button>	Clickable button
<select>	Drop-down list
<optgroup>	Options in a drop-down list
<option>	Option in a drop-down list
<label>	Label for an input element
<fieldset>	Groups related elements in a form
<legend>	Caption for a fieldset element

Forms:

It is an HTML form for user input.

Form Attributes:

- Action: It has a URL attached where the form data needs to be send
- Autocomplete: Turns on/off autocomplete option.
- Name: It gives name to a form.

```
<!doctype html>
<html>
<head>
<title>Training Page</title>
</head>
<Body>
<h1>Forms and Input</h1>
<p>How to Use Form Tag:</p>
<form action="demo" formtarget_top>
Name: <input type="text" value=""> <br> <br>
Phone: <input type="Numbers" value=""> <br> <br>
<input type="submit" value="submit">
</form>
</body>
</html>
```

OUTPUT: 1.05 – INTERNET EXPLORER:

Forms and input.html in Internet Explorer showing:

Forms and Input

How to Use Form Tag:

Name: [   ]

Phone: [   ]

[submit]

Text Area:

A text area is used to add unlimited characters.

Few attributes of Text area:

- cols :
  It is used to increase or decrease the width of the text area
- rows:
  It is used to increase or decrease the height of the text area,
- name :
  It is used to assign name to the text area

- required:

  It is an attribute that is used to make the user enter the details without fail.

- readonly:

  It is used to make the text area as read only. The user cannot alter the text content.

- wrap:

  This attribute is used to wrap the text when submitted.

EXAMPLE: TEXTAREA-1.06: NOTEPAD:

```
<!doctype html>
<html>
<head>
<title>Training Page</title>
</head>
<Body>
<h1>Forms and Input</h1>
<p>How to create a textarea:</p>
<textarea rows="5" cols="50">
This is a training module
</textarea><br>
<hr>
</body>
</html>
```

OUTPUT: 1.06 – INTERNET EXPLORER:

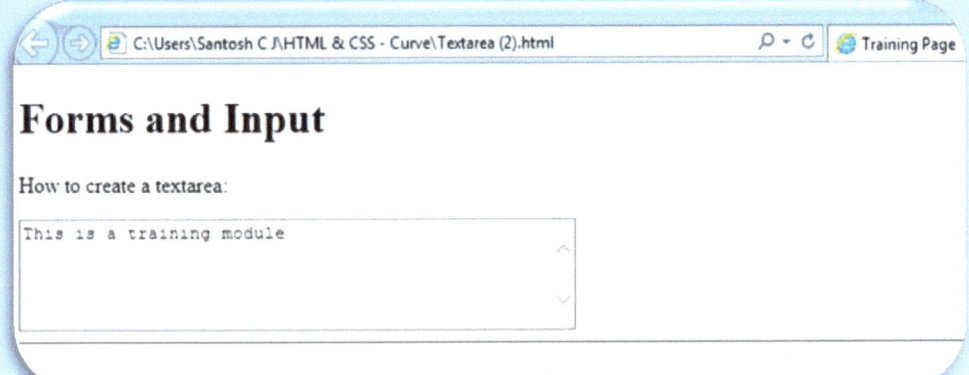

Buttons:

It is used to create a clickable button.

Button Attributes:

- Formaction: It has a URL attached where the form data needs to be send
- Formtarget: It specifies where to display the response content. It have different values like _self, _parent, _top etc.
- Type: It is show the type of button. It has different values like button, reset, submit etc.
- Value: It gives a value for the button.

EXAMPLE: BUTTONS-1.07: NOTEPAD:

```
<!doctype html>
<html>
<head>
<title>Training Page</title>
</head>
<Body>
<h1>Buttons</h1>
<p>How to create and to name a BUTTONS</p>
Search button: <button type="button"> Search </button>
<hr>
</body>
</html>
```

OUTPUT: 1.07 – INTERNET EXPLORER:

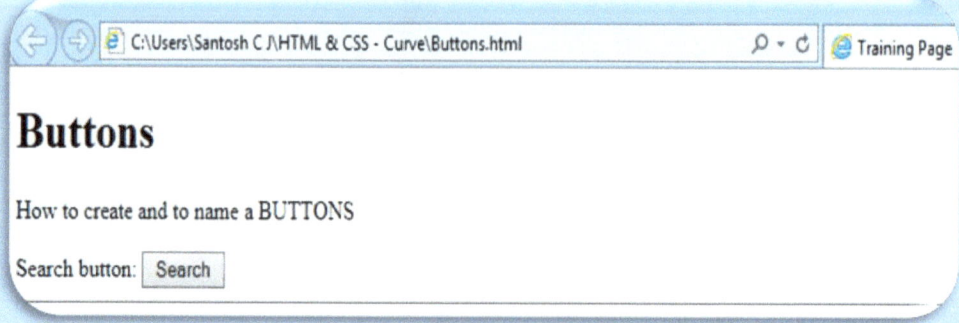

SELECT Tag:

It is used to create a drop down list.

Select attributes:

- Name: It gives name to the drop down list.
- Multiple: It allows the user to select multiple options.
- Size: It allows creating a number of visible options in the drop down list.

EXAMPLE: SELECT-1.08: NOTEPAD:

```
<!doctype html>
<html>
<head>
<title>Training Page</title>
</head>
<Body>
<h1>Select Tags</h1>
<p>How to create a select option</p>
<select>
<option value="Name">Name</option>
<option value="age"> Age</option>
<option value="Address">Address</option>
<option value="Pincode">Pincode</option>
</select>
<hr>
</body>
</html>
```

OUTPUT: 1.08 – INTERNET EXPLORER:

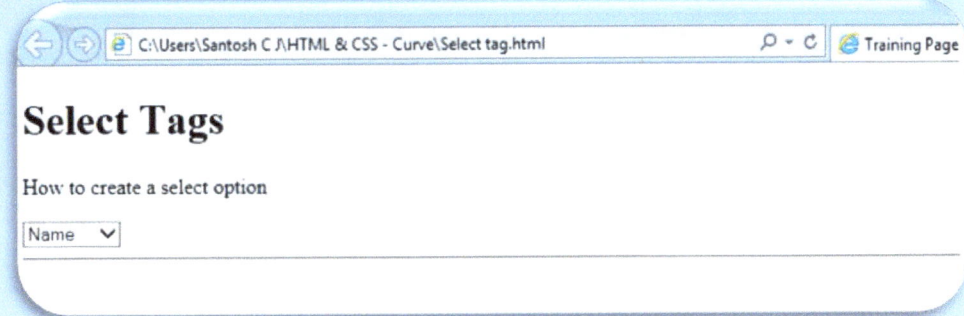

Optgroup Tag:

This tag is used to create related option list or a heading for a group of options. It is normally used with in the Select tag.

Optgroup Attributes:

- Label:

  It is used to give label to a specific group
- Disabled:

  It is used to disable a specific group in the option list.

EXAMPLE: OPTGROUP-1.09: NOTEPAD:

```html
<!doctype html>
<html>
<head>
<title>Training Page</title>
<meta name="Test" content="Basic Tags explanation"/>
</head>
<Body>
<h1>Select Tags</h1>
<p>How to create a select option</p>
<select>
<optgroup label="Personal info">
<option value="Name">Name</option>
<option value="age"> Age</option>
<option value="Address">Address</option>
<option value="Pincode">Pincode</option>
</optgroup>
<optgroup label="Acedemic info">
<option value="School">School</option>
<option value="College">College</option>
</optgroup>

</select>
<hr>
</body>
</html>
```

OUTPUT: 1.09 – INTERNET EXPLORER:

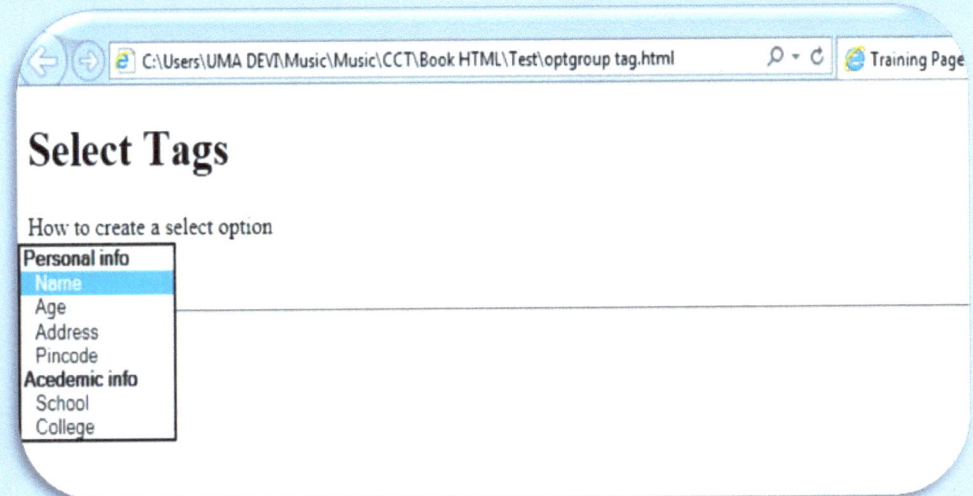

HT

You must be the change you wish to see in the world.

– Mahatma Gandhi

Label:

It helps us to create a Label for an input item.

Label Attributes:

- Form:
  This attribute helps us to know name a label for more than one form.
- For:
  This attribute is used to label a specific element in a form.

HTML & CSS - Curve

EXAMPLE: LABEL TAG-1.10: NOTEPAD:

```
<!doctype html>
<html>
<head>
<title>Training Page</title>
<meta name="Test" content="Tag explanation"/>
</head>
<Body>
<h1>Label Tags</h1>
<p>How to create a label</p>
<form action="demo">
<label="Looking for a bride">Looking for a bride:</label>
<select>
<option value="Yes"> YES</option>
<option value="No">No</option>
</select> <br> <br>
<label="Marital status"> Marital Status:</label>
<select>
<option value="Single">Single</option>
<option value="married">Married</option>
</select> <br> <br>
<input type="submit" value="submit">
</form>
<hr>
</body>
</html>
```

OUTPUT: 1.10 – INTERNET EXPLORER:

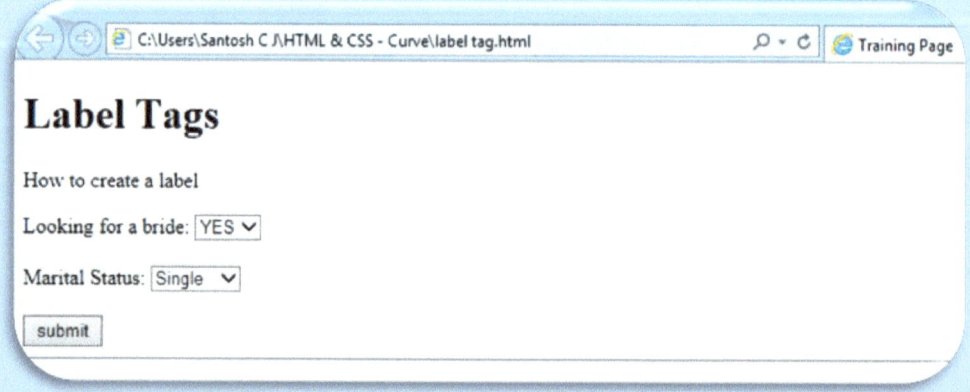

Fieldset:

It is used to create or group all related elements in side a box.

Attributes:

- Name:
  It gives name to a fieldset.
- Form:
  It helps the creator to assign one or more forms to a fieldset.

EXAMPLE: Fieldset TAG-1.11: NOTEPAD:

```
<!doctype html>
<html>
<head>
<title>Training Page</title>
<meta name="Test" content="Tag explanation"/>
</head>
<Body>
<h1>Label Tags</h1>
<p>How to create a label</p>
<form action="demo">
<fieldset>
<legend>Please select right option</legend>
<label="Looking for a bride">Looking for a bride:</label>
<select>
<option value="Yes"> YES</option>
<option value="No">No</option>
</select> <br> <br>
<label="Marital status"> Marital Status:</label>
<select>
<option value="Single">Single</option>
<option value="married">Married</option>
</select> <br> <br>
<input type="submit" value="submit">
</fieldset>
</form>
<hr>
</body>
</html>
```

OUTPUT: 1.11 – INTERNET EXPLORER:

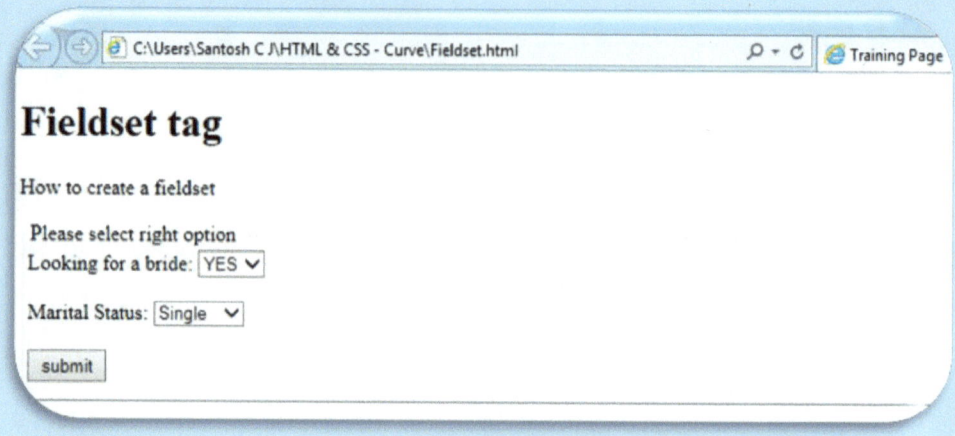

FRAMES:

HTML & CSS - Curve

FRAMES	
TAG	**DESCRIPTION**
<iframe>	It is an inline frame

iframe Attributes:

- Name:

This attribute gives name to the iframe.

- Src:

This attribute allows the creator to assign an URL to the iframe.

- height:

This attribute allows the designer to change the height of the frame

- width;

This attribute allows the designer to change the width of the frame.

EXAMPLE: IFRAME TAG-1.12: NOTEPAD:

```
<!doctype html>
<html>
<head>
<title>Training Page</title>
<meta name="Test" content="Tag explanation"/>
</head>
<Body>
<h1>Iframe tag</h1>
<p>How to create an iframe tag:</p>
<iframe src="https//www.google.com/" height="200px"
width="300px">
<hr>
</body>
</html>
```

OUTPUT: 1.12 – INTERNET EXPLORER:

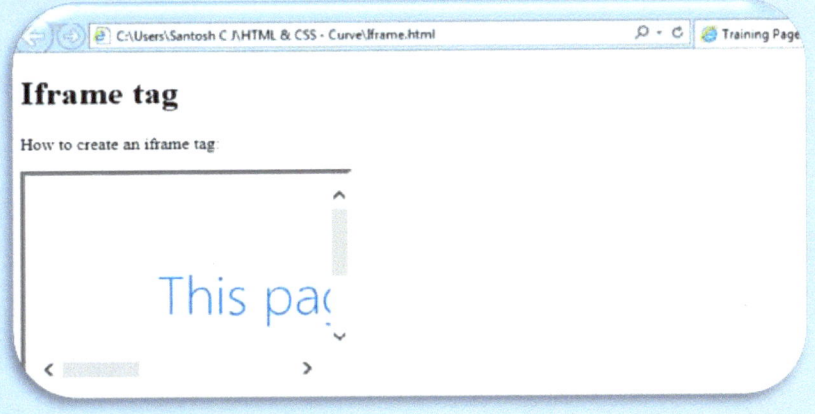

IMAGES:

The image tags are used to add Image components to the HTML document.

IMAGES	
Tag	**Description**
<img>	Defines an image

Img tag:

This tag is used to add a picture to the html document. The pictures can different extension like .png,, .jpg, .gif etc.

Attributes of img Tag:

- alt:

This attribute is used to add alternate text information to the image.

- Src:

This attribute allows the creator to assign an URL to the iframe.

- height:

This attribute allows the designer to change the height of the frame

- width;

This attribute allows the designer to change the width of the frame.

LOCATION OF THE PICTURE:

- If the image is the same folder where the html document is saved, then only the name of the picture with the extension is used like xyz.png, abc.jpg etc.

- If the image is not in the same folder, right click on the image and then choose properties, copy the location and use it in the HTML document along with its name and extension.

EXAMPLE: IMG-1.13: NOTEPAD:

```
<!doctype html>
<html>
<head>
<title>Training Page</title>
<meta name="Test" content="Tag explanation"/>
</head>
<Body>
<h1>Img tag</h1>
<p>How to use img tag:</p>
<img src="C:\Users\Santosh C J\pictures\abd.png"
alt="Symbol" width="200px" height="200px">
<hr>
</body>
</html>
```

OUTPUT: 1.13 – INTERNET EXPLORER:

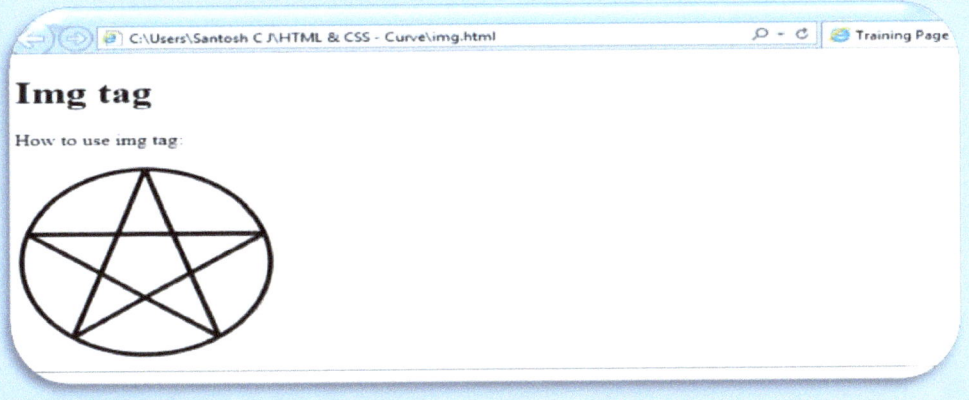

The desire of knowledge, like the thirst for riches, increases ever with the acquisition of it. – Laurence Sterne

AUDIO VIDEO:

This Audios and video tags allow the designer to add audio and video content links to the HTML document. Some of the frequently used Tags are as follows:

Audio / Video	
TAG	**DESCRIPTION**
<audio>	To add sound content
<video>	To add a video or movie

Audio tag:
This tag allows the web page designer to add Audio content links to the HTML document.

Audio tag Attributes:

- src:

This attribute is used add the URL of the audio file to the HTML document

- Controls:

This attribute is used to add audio control buttons to the HTML document. Some of the buttons include - Play/pause/ stop etc.

- muted:

This attribute allows the web designer to add mute option to the HTML document.

- autoplay:

This attribute allows the web designer to add an auto load option that makes the audio to play when the page loading completes.

- loop:

This attribute allows the web designer to add a loop option to the HTML document which enables the audio files to play in a loop.

EXAMPLE: AUDIO-1.14: NOTEPAD:

```
<!doctype html>
<html>
<head>
<title>Training Page</title>
<meta name="Test" content="Tag explanation"/>
</head>
<Body>
<h1>Audio Tag</h1>
<p>How to add an audio file:</p>
<audio controls>
<source src="C:\Users\Santosh C J\Music\bells.mp3"
type="audio/mp3">
</audio>
<hr>
</body>
</html>
```

OUTPUT: 1.14 – INTERNET EXPLORER:

Video Tag:

This tag is used to add video files to the HTML document.

Video Tag Attributes:

- src:

  This attribute is used add the URL of the audio file to the HTML document

- Controls:

This attribute is used to add audio control buttons to the HTML document. Some of the buttons include - Play/pause/ stop etc.

- muted:

This attribute allows the web designer to add mute option to the HTML document.

- autoplay:

This attribute allows the web designer to add an auto load option that makes the audio to play when the page loading completes.

- loop:

This attribute allows the web designer to add a loop option to the HTML document which enables the audio files to play in a loop.

EXAMPLE: VIDEO-1.15: NOTEPAD:

```
<!doctype html>
<html>
<head>
<title>Training Page</title>
<meta name="Test" content="Tags explanation"/>
</head>
<Body>
<h1>Video Tag</h1>
<p>How to add a video file:</p>
<video width="420" height="300" controls>
<source src="C:\Users\Santosh C J\Videos\2012-07\SO much.mp4"
type="video/mp4">
</video>
<hr>
</body>
</html>
```

OUTPUT: 1.15 – INTERNET EXPLORER:

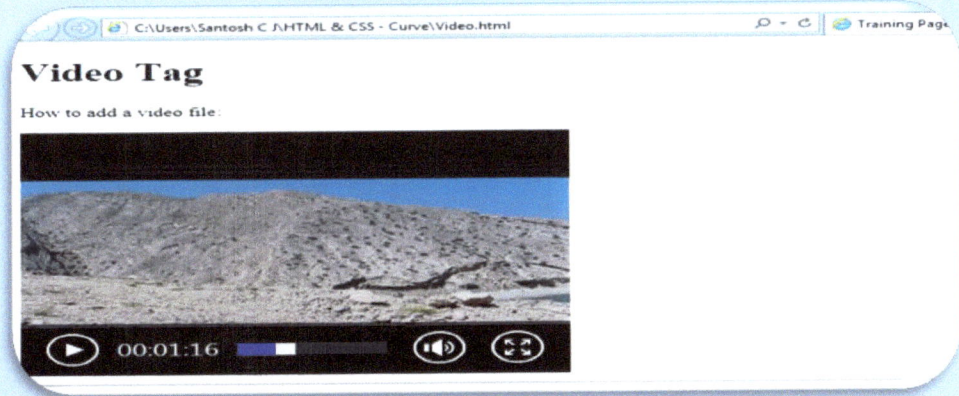

LINKS:

A link tag is used to add a Hyperlink to the html document. For example –
http://www.yahoo.com/.

LINKS	
TAG	**DESCRIPTION**
<a>	To add a hyperlink

EXAMPLE: LINK TAG-1.16: NOTEPAD:

```
<!doctype html>
<html>
<head>
<title>Training Page</title>
<meta name="Test" content="Tags explanation"/>
</head>
<Body>
<h1>Link Tag</h1>
<p>How to add a Hyperlink to a HTML file:</p>
<P>Please click on the link below</p>
<a href="http://www.hotmail.com/"> [Hotmail]</a>
<hr>
</body>
</html>
```

OUTPUT: 1.16 – INTERNET EXPLORER:

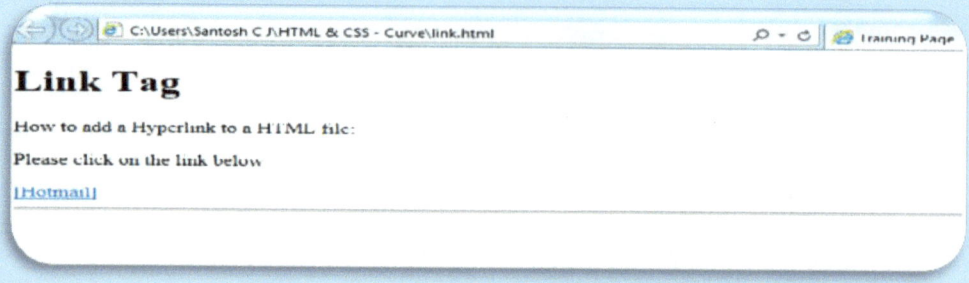

LISTS:

A list tag is used to create Lists in a web page. Few list tags:

LISTS	
TAG	**DESCRIPTION**
<ul>	Unordered list
<ol>	Ordered list
<li>	List item
<dl>	Description list

EXAMPLE: LIST TAG-1.17: NOTEPAD:

```
<!doctype html>
<html>
<head>
<title>Training Page</title>
</head>
<Body>
<h1>Lists Tag</h1>
<h1> Tips to live a healthy life</h1>
<p><ul>
<li>Wake up early.</li>
<li>Eat healthy food.</li>
</ul>
<h2>Do and Don'ts</h2>
<p><ol>
<dl>1) Never fight with your neightbour</dl>
<dl>2) Never lie.</dl>
</ol></p>
<hr>
</body>
</html>
```

OUTPUT: 1.17 – INTERNET EXPLORER:

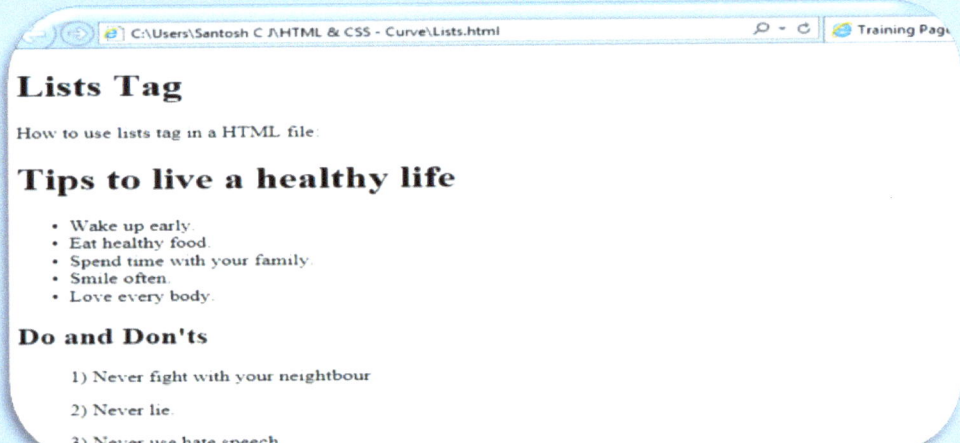

TABLES:

The table tags are used to draw tables in a HTML document. Some of the tags are:

TABLES	
TAG	**DESCRIPTION**
<table>	Table
<caption>	Caption for the table
<th>	header of a table
<tr>	Row in a table
<td>	Cell in a table
<thead>	Header content
<tbody>	Body content
<tfoot>	Foot content

I fail. My God doesn't.

Table Tag: EXAMPLE: Table tag-1.18: NOTEPAD:

```html
<!doctype html>
<html>
<head>
<title>Training Page</title>
<style>
table, td, th
{
border:1px solid red;
}
</style>
</head>

<Body>
<h1>Table Tags:</h1>

<table>
<caption> Schedules</caption>

<thead>
<tr>
<th>S.NO</th>
<th>Name of the commodity</th>
<th>Price in Rs</th>
</tr>
</thead>

<tbody>
<tr>
<td>1)</td>
<td>Noodles</td>
<td>30</td>
</tr>
<tr>
<td>2)</td>
<td>Chocolate</td>
<td>20</td>
<tr>
</tbody>

<tfoot>
<tr>
<td>[+]</td>
<td>Total</td>
<td>50</td>
</tr>
</tfoot>
</table>
</body>
</html>
```

OUTPUT: 1.18 – INTERNET EXPLORER:

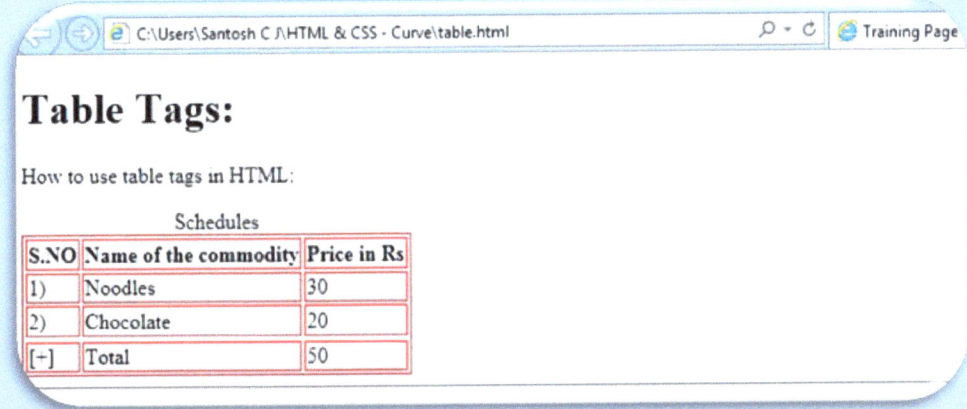

HTML Background:

Backgrounds:

The <body> tag has two attributes where you can specify backgrounds. The background can be a color or an image.

Background Image:

The background attribute can also specify a background-image for an HTML page. The value of this attribute is the URL of the image you want to use. If the image is smaller than the browser window, the image will repeat itself until it fills the entire browser window.

EXAMPLE: COLOR CODES-1.19: NOTEPAD:

```html
<!doctype html>
<html>
<head>
<title>Training Page</title>
<meta name="Test" content="About HTML Background"/>
</head>
<Body bgcolor="skyblue">

<h1>HTML Background</h1>
<p>How to add colors to a HTML file:</p>
</body>
</html>
```

Bgcolor:

The bgcolor attribute specifies a background-color for an HTML page. The value of this attribute can be a hexadecimal number, an RGB value, or a color name:
<body bgcolor="#000000"> <body bgcolor="rgb (0,0,0)"> <body bgcolor="black">
The lines above all set the background-color to black.

Color Values

Colors are defined using a hexadecimal notation for the combination of red, green, and blue color values (RGB). The lowest value that can be given to one light source is 0 (hex #00). The highest value is 255 (hex #FF).

EXAMPLE: COLOR CODES-1.20:

black (#000000)	silver (#C0C0C0)	gray (#808080)	white (#FFFFFF)
maroon (#800000)	red (#FF0000)	purple (#800080)	fuchsia (#FF00FF)
green (#008000)	lime (#00FF00)	olive (#808000)	yellow (#FFFF00)
navy (#000080)	blue (#0000FF)	teal (#008080)	aqua (#00FFFF)

EXAMPLE: COLOR CODES-1.21: NOTEPAD:

```
<!doctype html>
<html>
<head>
<title>Training Page</title>
<meta name="Test" content="About HTML Background"/>
</head>
<Body background="C:\Users\Santosh C J\pictures\oxygen.jpg">
<h1>HTML Background</h1>
<p>How to add a background picture to a HTML file:</p>
</body>
</html>
```

HTML & CSS - Curve

CSS

CSS - Cascading Style Sheets is a style sheet used in HTML language. It is used for creating and formatting the content of a document written in a markup language.

USES:

1) Helps to separate presentation from content
2) Helps in giving same look to all the pages
3) Helps in designing the font, font- size, color and other properties of displayed text
4) Helps in designing the background pictures and Colors.
5) Easy way to control the appearance of web sites.

Basic Syntax:

CSS has a simple syntax and uses a number of English keywords to specify the names of various style properties.

A style sheet consists of a list of rules. Each rule or rule-set consists of one or more selectors, and a declaration block.

Selector {Property: Value}

Basic rules:

1) Each selector can have multiple properties.
2) Each property within that selector can have independent values.
3) The property and value are separated with a colon contained within curly brackets.
4) Multiple properties are separated by a semi colon.
5) Multiple values within a property are separated by commas, and if an individual value contains more than one word you surround it with quotation marks.

Inline, Internal and external Style sheets:

A CSS or Cascading style sheet helps the designer to separate the HTML content from its style. The designer uses the HTML content to create a component and to arrange the component. The designer will use the CSS to alter the fonts, size, color, borders, formatting, link effects etc. The CSS components can either be placed internally or externally.

Inline style sheet:

In this method the designer uses the CSS cods alongside the element.

<h1 style="font-size:12px; font-family: Arabic typesetting; text-decoration:underline">

Internal Style sheet:

In this method the CSS code is completely placed inside the <head> tag of the HTML document. This method is useful only when the designer wants to design a single page.

```
<!doctype html>
<html>
<head>
<title><title>
<style type="text/css">
CSS Content Goes Here
</style>
</head>
<body>

</body>

</html>
```

External Style sheet:

In this method the CSS code is written using a notepad or Text editor and then the File is saved with a .CSS extension. This external file is linked to the HTML document. It is normally added to the <head></head> section.

<link rel="stylesheet" type="text/css" href=*"Path To stylesheet.css"* />

DIVISIONS:

Divisions are a block level HTML element used to define sections of a HTML file. It contains all the parts that is used to create a website. The division part is placed in between <body></body> tags. A designer can have multiple divisions inside the <body></body> tag.

```
<!doctype html>

<html>

<head>

<title></title>

<style>

</style>

<body>

<div>

<h1></h1>

</div>

</body>

</html>
```

CSS IDs:

IDs are similar to classes, except once a specific id has been declared it cannot be used again within the same HTML file.

```
<div id="Example">
--Contents--
</div>
```

CSS file:

```
<style>

#example{

Font-size:20px;

Color:red;

Text-decoration:underline;

}

</style>
```

When my absence doesn't alter your life, then my presence has no meaning in it.

EXAMPLE: CSS EXAMPLE-2.01: NOTEPAD:

```
<!doctype html>
<html>
<head>
<title>Training Page</title>

<style>
#Example
{
font-size:40px;
color:red;
background:skyblue;
}

</style>
</head>
<body>

<div id="Example">
CSS EXAMPLE
</div>

</body>
</html>
```

OUTPUT: 2.01 – INTERNET EXPLORER:

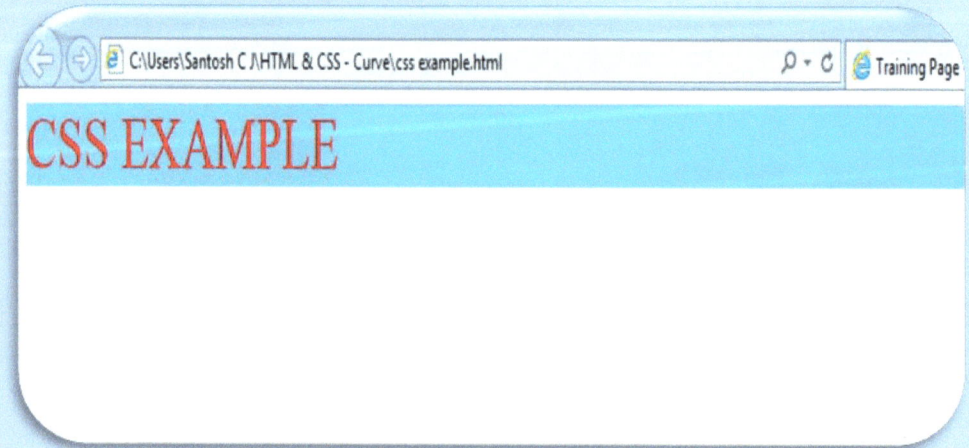

FONTS - CSS:

In CSS there are different ways to Style fonts. They are classified in to

- **font-family -** Face of a font.
- **font-style**- to make a font italic or oblique.
- **font-variant** used to create a small-caps effect.
- **font-weight -** bold or light a font.
- **font-size -** increase or decrease the size of a font.

Font-family:

The CSS font family can be classified in to three groups:

1) Sans serif
2) Serif
3) Monospaced

Sans Serif:

- o Arial
- o Arial Black
- o Arial Narrow
- o Arial Rounded MT Bold
- o Avant Garde
- o Calibri
- o Candara
- o Century Gothic
- o Franklin Gothic Medium
- o Futura
- o Geneva
- o Gill Sans
- o Helvetica
- o Impact
- o Lucida Grande
- o Optima
- o Segoe UI
- o Tahoma
- o Trebuchet MS
- o Verdana

Serif:

- o Baskerville
- o Big Caslon
- o Bodoni MT
- o Book ANtiqua
- o Calisto MT
- o Cambria
- o Didot
- o Garamond
- o Georgia
- o Goudy Old Style
- o Hoefler Text
- o Lucida Bright
- o Palatino
- o Perpetua
- o Rockwell
- o Rockwell Extra Bold
- o Times New Roman

Monospaced:

- o Andale Mono
- o Consolas
- o Courier New
- o Lucida Console
- o Lucida Sans Typewriter
- o Monaco

Fantasy:

- o Copperplate
- o Papyrus

Script:

- Brush Script MT

EXAMPLE: FONT-FAMILY-2.02: NOTEPAD:

```
<!doctype html>
<html>
<head>
<title>Training Page</title>
<meta name="CSS" Content="Explanation about CSS FONTS"/>
<style>
body{
background:skyblue;
}
#futura{
font-family:futura;
font-style:italic;
font-variant:small-caps;
font-weight:bold;
font-size:40px;
}
#impact{
font-family:impact;
font-style:oblique;
font-variant:none;
font-weight:light;
font-size:50px;
}
#fantasy{
font-family:fantasy;
font-style:none;
font-variant:none;
font-weight:none;
font-size:60px;
}
</style>

<body>
<div id=futura>
CSS FONT-FAMILY - Futura
</div>

<div id=impact>
CSS FONT-FAMILY - impact
</div>

<div id=fantasy>
CSS FONT-FAMILY - Papyrus
</div>

</body>
</html>
```

HTML & CSS - Curve

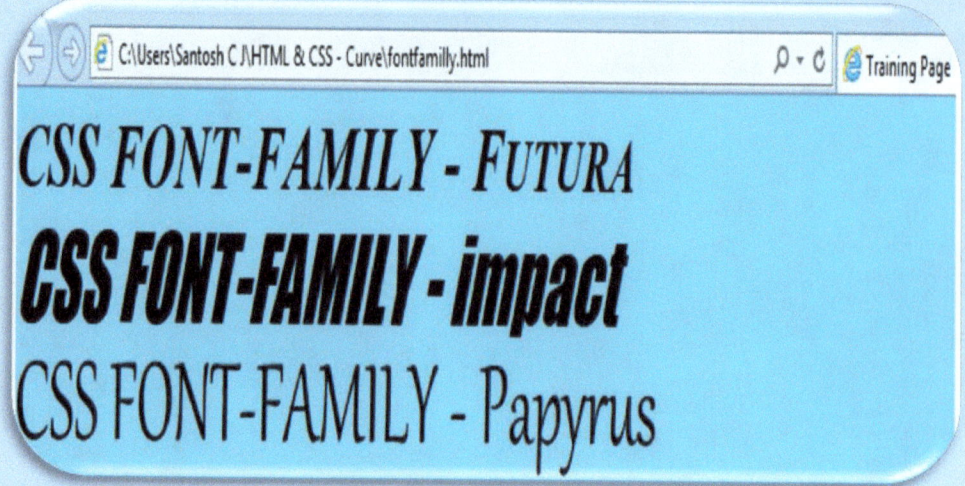

CSS Text Properties:

- Color
- Letter spacing
- Text align
- Text decoration
- Text indent
- Text Transform
- White space
- Word spacing

Color:

It is used to change the color of the text. The value can be Red, blue, yellow, olive etc.

color: value;

Letter Spacing:

It is used to add space between letters. Values are 1) Normal & 2) High.

letter-spacing: value;

Text Align:

This is used to align the text to different locations of the web page. It can be given values like – left, right, center and justify.

text-align: value;

Text Decoration:

This is used to decorate the text. It can be given values like – none, underline, link through, overline and blink.

Text-decoration: value;

Text indent:

This is used to indent the first line of the text in an HTML document. It can be given values like length and percentage.

text-indent: value;

Text Transform:

This is used to convert the text in to capital or lowercase. It can be given values like none, capitalize and lowercase.

text-transform: value;

Word spacing:

This is used to adjust space between words. It can be given values like Normal and length.

word-spacing: value;

The only way around is through. – Robert Frost

EXAMPLE: CSS TEXT PROPERTIES-2.03: NOTEPAD:

```
<!doctype html>
<html>
<head>
<title>Training Page</title>
<meta name="CSS" Content="Explanation about CSS Properties"/>
<style>
body {
background skyblue;
font-size 40px;
}

#first {
color red;
letter-spacing 5px;
text-align left;
text-decoration underline;
text-indent 10px;
text-transform uppercase;
word-spacing 20px;
}

#second {
color blue;
letter-spacing 10px;
text-align center;
text-decoration none;
text-indent 20px;
text-transform lowercase;
word-spacing 30px;
}

#third {
color yellow;
letter-spacing 15px;
text-align right;
text-decoration overline;
text-indent 30px;
text-transform none;
word-spacing 5px;
}
</style>
<body>
<div id=first>
Text Decoration - first
</div>
<div id=second>
Text decoration - second
</div>
<div id=third>
Text decoration - third
</div>
</body>
</html>
```

OUTPUT: 2.03 – INTERNET EXPLORER:

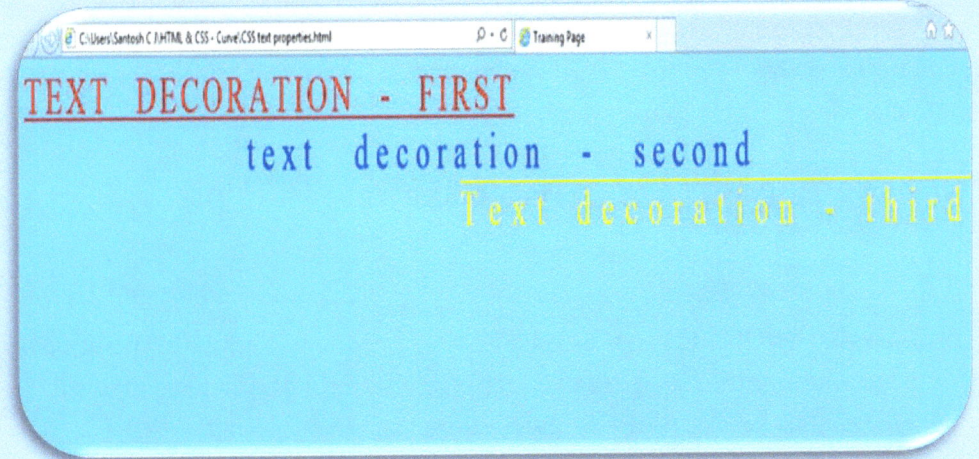

CSS BACKGROUND:

The background property is used to design the HTML document. It allows the designer to add or change background image, color, attachment etc. It can be given different values like

- color
- image
- Attachment
- Position
- Repeat

Color:

It is used to add a background color to the HTML document. Colors like Red, blue, green, yellow etc., can be added.

Background-color: value;

Image:

It is used to add an image as a background to the HTML file. Right click on the picture and choose properties. Use the location or the URL in order to add an image. Most of the extensions are used like .JPG, .GIF, .PNG etc.

Background-image: URL (location of the picture);

Attachment:

It is used, when the designer is using an image as the background. The designer can set whether the background Image scrolls along with the page or it is fixed when the user scrolls. It can be given values like scroll or fixed.

Background-attachment: value;

Position:

It is used to position an image which is used as the background of an HTML document. It can be given different values like top left, top center, top right, center left, center center, center right, bottom left, bottom center, bottom right, x-% y-% and x-pos y-pos.

Background-position: value;

Repeat:

It is used to set if an image set as a background of an element is to repeat (across=x and/or down=y) the screen using the background-repeat property.

Background-repeat: value;

You must remain focused on your journey to greatness. – Les Brown

EXAMPLE: CSS BACKGROUND COLOR-2.04: NOTEPAD:

```
<!doctype html>
<html>
<head>
<title>Training Page</title>
<meta name="CSS" Content="Explanation about CSS Properties"/>
<style>

body{
background-color:skyblue;
}
</style>

<body>

<div>
<h1>Welcome to Sam Studios 96</h1>
</div>

</body>
</html>
```

OUTPUT: 2.04 – INTERNET EXPLORER:

EXAMPLE: CSS BACKGROUND IMAGE-2.05: NOTEPAD:

NOTE: The Picture which is used as the background Image has to be saved in the folder where the HTML file is saved.

```html
<!doctype html>
<html>
<head>
<title>Training Page</title>
<meta name="CSS" Content="Explanation about CSS Properties"/>
<style>
body {
background-image:url('oxygen.jpg');
background-position:bottom center;
background-attachement:fixed;
}
</style>
<body>
<div>
<h1>Welcome to Sam Studios 96</h1>
</div>
</body>
</html>
```

OUTPUT: 2.05 – INTERNET EXPLORER:

CSS BORDER:

The border property is used to adjust the color, style and width of the borders. Some of the values used are

- Color
- Style
- Width.

border: 1px solid value;

Border color:

It is used to change the color of the border. Some of the values frequently used are Color name, hexadecimal number, RGB Color Code, transparent.

border-color: value;

Border style:

It is used to change the style of the border. Some of the values frequently used are dashed, dotted, double, groove, hidden, inset, none, outset, ridge and solid.

border-style: value;

Border width:

It is used to adjust the width of the border. Some of the values frequently used are length, thin, medium and thick.

border-width: value;

or Each border side can be changed or altered individually.

Example:

Border-bottom: 1px solid red;

Border-left-style: groove;

EXAMPLE: CSS BORDER-2.06: NOTEPAD:

```
<!doctype html>
<html>
<head>
<title>Training Page</title>
<meta name="CSS" Content="Explanation about CSS Properties"/>
<style>
body{
background-color:skyblue;
}
h1{
border:2px solid brown;
border-style:dashed;
border-width:5px;
border-left:3px solid blue;
}
h2{
border:2px solid yellow;
border-style:groove;
border-bottom:2px solid red;
}
</style>

<body>
<div>
<h1>Border Example 1</h1>
<h2>Border Example 2</h2>
</div>
</body>
</html>
```

OUTPUT: 2.06 – INTERNET EXPLORER:

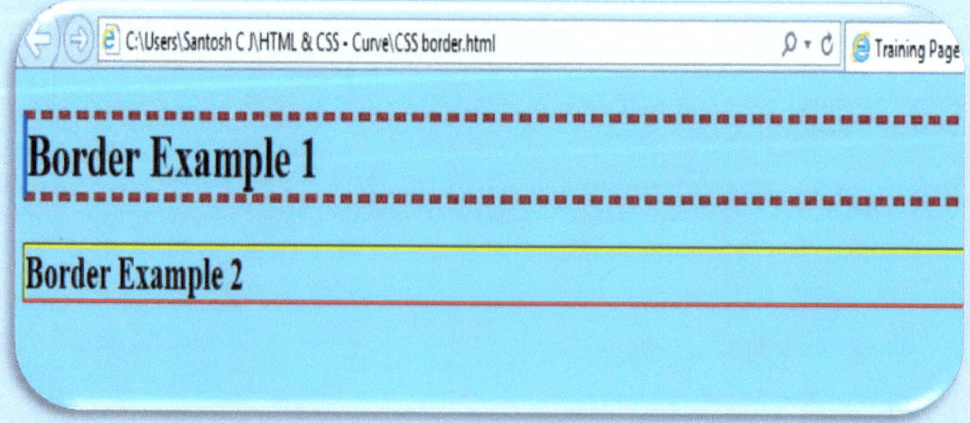

CSS LIST STYLE:

This is used to adjust the appearances of ordered and unordered lists. Some of the values frequently used are

- Image,
- Position
- Type.

List-type: value value;

List style Image:

It is used to add image for the bullet of unordered lists.

List-style-image: url ('path to image');

List style Position:

It is used to control the position of the unordered and ordered lists. Some of the values frequently used are inside and outside.

List-style-position: value;

List style type:

It is used to change the type of bullet used with the ordered or unordered lists. Some of the values frequently used are disc, circle, square, decimal, lower-roman, upper-roman, lower- alpha, upper-alpha and none.

List-style-type: value;

> You must either modify your dreams or magnify your skills. – Jim Rohn

EXAMPLE: CSS BORDER-2.07: NOTEPAD:

```
<!doctype html>
<html>
<head>
<title>Training Page</title>
<meta name="CSS" Content="Explanation about CSS Properties"/>

<style>
body {
background-color skyblue;
}
#Dos {
list-style-position inside;
list-style-type lower-roman;
}
#health {
list-style-position outside;
list-style-image url('egg 4.jpg');
}
</style>

<body>

<div id=Dos>
<caption>DOs and Don'ts</caption>
<ul>
<li>Wake up early</li>
<li>Brush your teeth twice a day</li>
<li>You must take your break fast before 8 AM</li>
<li>You must go to bed early</li>
</ul>
</div>

<div id=health>
<caption>Healthy habits</caption>
<ul>
<li>Take bath daily</li>
<li>Eat less oily food</li>
<li>Practise Yoga</li>
<li>Exercise atleast for 30mins</li.
</ul>
</div>

</body>
</html>
```

<u>OUTPUT: 2.07 – INTERNET EXPLORER:</u>

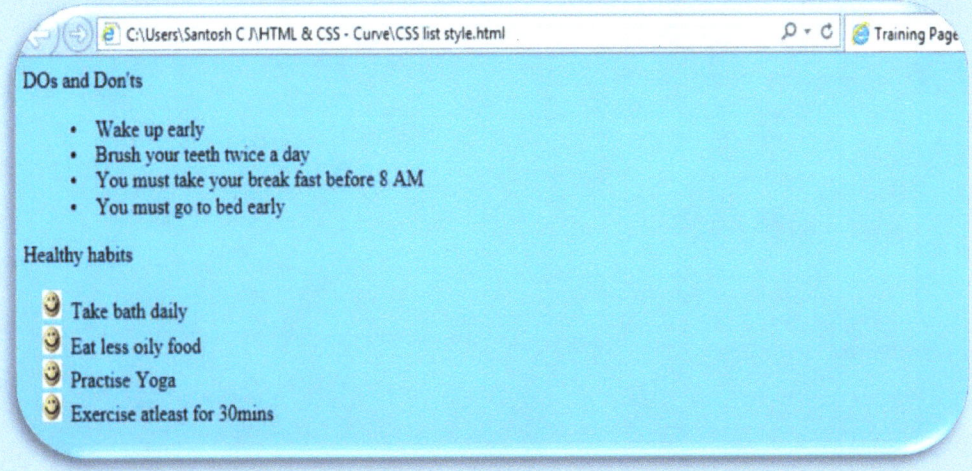

CSS ANCHORS & LINKS:

It is used to change the appearances of the links created in a HTML document. Some of the ways to change the appearances are:

a:link {color: red;} – Sets color to the link.

a:visited {color: blue;} – Changes color when user has visited the page by clicking on the link.

a:hover {color: olive;} – Changes color when user places mouse on the link.

a:focus {color: green;} – Changes color when user places mouse or use TAB to navigate.

a:active {color: black;} – Changes color when user presses the mouse button on the link.

<u>EXAMPLE: CSS LINKS-2.08: NOTEPAD:</u>

```html
<!doctype html>
<html>
<head>
<title>Training Page</title>
<meta name="CSS" Content="Explanation about CSS Properties"/>
<style>
body{
background-color:skyblue;
}
#link{
border:2px solid red;
font-size:20px;
}
a:link{
color:Blue;
}
a:hover{
color:yellow;
}
</style>
<body>

<div id=link>
<p>Click on the link below</p>
<a href="www.google.com/">Google</a>

</div>
</body>
</html>
```

<u>OUTPUT: 2.08 – INTERNET EXPLORER:</u>

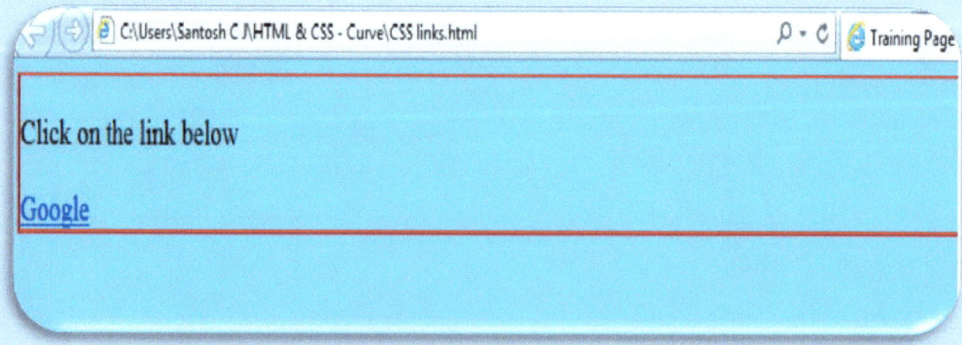

Other CSS PROPERTIES:

There are other few properties which are used frequently by designers. Some of them are:

1. Margin
2. Padding
3. Height
4. Width
5. Clear
6. Clip
7. Cursor
8. Display
9. Float
10. Overflow
11. Position
12. Span.

1) CSS MARGIN:

It is used to create margin between the HTML element and the other elements around it.

margin-top: length percentage or auto;
margin-left: length percentage or auto;
margin-right: length percentage or auto;
margin-bottom: length percentage or auto;

For example:

Margin:5px 5px 5px 5px;

It assigns a length of 5px to top, right, left and bottom of the margin.

2) *CSS PADDING:*

Padding is the distance between the border of an HTML element and the content within it. Some of the values frequently used are length and percentage.

padding-top: length percentage;
padding-left: length percentage;
padding-right: length percentage;
padding-bottom: length percentage;

For example:

Padding: 5px 5px 5px 5px;

It assigns a value of 5px for padding on left, right, top and bottom.

3) *CSS HEIGHT:*

It is used to control the Height of different elements in an HTML document. Some of the frequently used values are Auto, length and percentage.

height: value;

`Different height Components used in CSS:
- Line height
- Max height
- Min height

Line height:

It is used to increase the height between the lines in a HTML document. Some of the values are normal, number, length and percentage.

Line-height: value;

Max height:

It is used to control the maximum height of an element. Some of the frequently used values are none, length and percentage.

Max-height: value;

Min height;

It is used to control the minimum height of the element. Some of the frequently used values are length and percentage..

Min-height: value;

4) *CSS WIDTH:*

It is used to control the width of an element. Some of the frequently used values are auto, length and percentage.

Width: value;

Different width components used in CSS:

- Max width
- Min width

Max width:

It used to control the maximum width of the element in a HTML document. Some of the frequently used values are none, length and percentage.

Max-width: value;

Min width:

It is used to control the Minimum width of the element in a document. Some of the frequently used values are length and percentage.

Min-width: value;

5) *CSS CLEAR:*

It is used to control if an element allows floated elements to its sides. Some of its values are

- None – Default setting, floated elements can appear on either side.
- Both – It clears floated element on both the sides
- Left – No floated elements to appear on the left side.
- Right – no floated element to appear on the right side.

Clear: value;

6) *CSS CLIP:*

It allows the designer to control the visibility of the element. Some of the frequently values are auto and shape.

Clip: value;

HTML & CSS - Curve

7) *CSS Cursor:*

It is used to control the style of a cursor. Some of the frequently used values are auto, crosshair, default, help, move, pointer, text, url, wait, e-resize, ne-resize, nw-resize, n-resize, se-resize, sw-resize, s-resize and w-resize.

Cursor: value;

8) *CSS DISPLAY:*

It is used control the display properties allowing the designer to change the way an element looks. Some of the values are

- Block – Line break before and after the element
- Inline – No line break
- List item – Adds a list with the line break
- None – It doesn't allow the element to be displayed.

Display: value;

9) *CSS FLOAT:*

It helps the designer to make changes the text or image with in an element. Some of the frequently used values are left, right, none etc.

Float: value;

10)   *Overflow:*

The overflow property is used to control the contents of a box when it overflows. Some of the frequently used frequently used values are

- Auto
- Hidden
- Visible
- Scroll

Overflow: value;

HTML & CSS - Curve

11)   *Position:*

It is used to position the elements in a HTML document. Some of the values frequently used are

- Static - Default position of the element.
- Relative- Relatively placed with regards to other elements in the page
- Absolute – Positions it to top left of the nearest element.
- Fixed – It doesn't make any difference.

Position: value;

12)   *CSS SPAN:*

It is an inline element. It is just like Division tag<div> </div> which handles blocks of elements. Span can be used between texts to makes necessary changes.

```
<span class="test">Sample</span>
```

CSS Span:

```
.test {Color:red;}
```

Some people come into our lives, leave footprints on our hearts, and we are never the same.

SAMPLE DESIGN

Sample 1: Design – Home page:

```html
<!doctype html>
<html>
<head>
<title>Design 1</title>
<meta name="Design" content="Sample web site design"/>
<style type="text/css">

body
{
background-color:gray;

}

h1{
color:black;
font-size:60px;
font-family:arabic typesetting;
padding-left:20px;
background-color:green;
text-decoration:underline;
}

#box {
border:1px solid black;
height:50px;
background-color:brown;
text-align:center;
font-size:20px;
padding-bottom:25px;
}


A:link{
color:blue;
}
A:hover{
color:black;
}
A:visited{
color:silver;
```

```
}

#search{
padding-left:136px;
position:absolute;
margin-top:-20px;
}


div.img {
border:1px solid black;
    margin: 5px;
    padding: 5px;
    border: 1px solid #0000ff;
    height: auto;
    width: auto;
    float: left;
    text-align: center;
}

div.img img {
    display: inline;
    margin: 5px;
    border: 1px solid #ffffff;
}

div.img a:hover img {
    border:1px solid #0000ff;
}

div.desc {
    text-align: center;
    font-weight: normal;
    width: 120px;
    margin: 5px;
}


</style>


<body>


<div style="padding-left:910px; cursor:pointer">

<h7>
<a href="https://www.xyz.com/Myaccount/">My account,</a>
<a href="https://www.xyz.com/login/">Login,</a>
```

```
</h7>

<form id=search>
<input type="text" value="Search web">
</form>

</div>


<div>
<h1>Welcome to Sam Studios</h1>
<label style="position:absolute; left:925px; margin-top:-79px">I
am</label>
<select style="Position:absolute; left:963px; margin-top:-79px">
<option value="visiting for the first time">Visiting for the
first time.</option>
<option value="new to this website">new to this web
site.</option>
<option value="a member">a member.</option>
</select>
<div style="position:absolute; left:1137px; margin-top:-80px">
<button type="button">Submit</button>
</div>
</div>

<div id=box>

<h2>
<a href="https://www.xyz.com/home">[ HOME ]</a> <a
href="https://www.xyz.com/products/">[ PRODUCTS ]</a> <a
href="https://www.xyz.com/support/">[ SUPPORT ]</a> <a
href="https://www.xyz.com/contact/">[ CONTACT ]</a></h2>

</div>

<div style="border:1px solid red; padding-left:263px;
background:maroon;">


<img src="1.jpg" alt="background pic main page" width="800px"
height="400px">
</div>

<div id=box2>
<h3 style="text-decoration:underline">Introduction to Sam
Studios:</h3>
<p style="font-family:Arabic typesetting; font-size:22px"> Sam
Studios was founded be XYZ. It was established in the year
```

```
AA/BB/CCCC. It has branches across the globe. It is one of the
leading publishers with many established authors as its member.
<br> We also provide assistance in Music and Movie production. It
has links across the globe.  </p>
</div>


<hr>



<div style="border:1px solid black; height:300px;
background:teal; padding-left:55px">
<h6 style="text-decoration:underline; text-align:center; font-
size:18px; font-family:impact; color:maroon">Image gallery</h6>
<div class="img">

  <a target="_blank" href="a.htm">
    <img src="a.jpg" alt="a" width="110" height="90">
  </a>
  <div class="desc">Cat and ducks</div>
</div>
<div class="img">
  <a target="_blank" href="b.htm">
    <img src="b.jpg" alt="b" width="110" height="90">
  </a>
  <div class="desc">Cute Cat</div>
</div>
<div class="img">
  <a target="_blank" href="c.htm">
    <img src="c.jpg" alt="c" width="110" height="90">
  </a>
  <div class="desc">Dark Angel</div>
</div>
<div class="img">
  <a target="_blank" href="d.htm">
    <img src="d.jpg" alt="d" width="110" height="90">
  </a>
  <div class="desc">Blind Angel</div>
</div>
<div class="img">
  <a target="_blank" href="e.htm">
    <img src="e.jpg" alt="e" width="110" height="90">
  </a>
  <div class="desc">Lonely Angel</div>
</div>
<div class="img">
  <a target="_blank" href="f.htm">
    <img src="f.jpg" alt="f" width="110" height="90">
  </a>
```

```html
    <div class="desc">Unique Angel</div>
</div>
<div class="img">
  <a target="_blank" href="g.htm">
    <img src="g.jpg" alt="g" width="110" height="90">
  </a>
  <div class="desc">Arc Angel</div>
</div>
<div class="img">x
  <a target="_blank" href="h.htm">
    <img src="h.jpg" alt="h" width="110" height="90">
  </a>
  <div class="desc">Warrior Angel</div>
</div>


</div>



<hr>
<div>
<h4 style="text-align:center; text-decoration:underline;
color:Red; font-family:Arial">Disclaimer</h4>
<p style="text-align:center; font-family:TIMES NEW ROMAN">This
web site and its content is a copyright of <a
href="https//www.xyz.com/home/">www.xyz.com</a>. All rights
reserved.<br>

Any redistribution or reproduction of part or all of the contents
in any form is prohibited. You may not, except with our express
written permission, distribute or commercially exploit the
content. Nor may you transmit it or store it in any other website
or other form of electronic retrieval system.<br>

The web site and its content may not be copied, reproduced,
republished, downloaded, posted, broadcast or transmitted in any
way without first obtaining <b>XYZ</b> written permission or that
of the copyright owner. For more information contact through
<b>Email:</b> <i>info@xyz.com</i></p><hr>
</div>


</body>
</html>
```

```
============================================================
```

OUTPUT: Sample:1 – INTERNET EXPLORER:

My account, Login Search web

Welcome to Sam Studios

I am Visiting for the first time. ∨   Submit

[HOME] [PRODUCTS] [SUPPORT] [CONTACT]

Introduction to Sam Studios:

Sam Studios was founded by XYZ. It was established in the year AA/BB/CCCC. It has branches across the globe. It is one of the leading publishers with many established authors as its member. We also provide assistance in Music and Movie production. It has links across the globe.

Image Gallery

  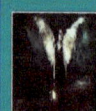     

Cat and ducks   Cute Cat   Dark Angel   Blind Angel   Lonely Angel   Unique Angel   Arc Angel   Warrior Angel

Disclaimer

Sample 2: Design – Home page:

```
<!doctype html>
<html>
<head>
<title>Design form</title>
<meta name="Design2" content="Sample web site design"/>
<style type="text/css">

body{
background:green;
}


a:hover{
color:red;
}
a:visited{
color:black;
}

#box1{
border:1px solid maroon;
width:200px;
height:1500px;
position:absolute;
left:1100px;
background:maroon;
}

#box2{
border:1px solid maroon;
width:20px;
height:1500px;

background:maroon;
}


a:link{
hover:red;
}
a:visited{
color:black;
}
```

```
</style>


<body>

<div id=box1>

</div>

<div id=box2>
</div>


<div style="position:absolute; left:60px; margin-top:-1500px;
text-decoration:underline; color:white; font-family:arabic
typesetting; font-size:60px; border:1px solid skyblue;
width:940px; padding-left:70px;background:skyblue">
<h1>Government Of India</h1>

</div>

<div id=rom>
<div style="position:absolute; left:710px; margin-top:-1280px">
<p><a href="https://www.goi.com/myaccount/">My account,</a> <a
href="https://www.goi.com/login"/>Login,</a></p>
<form style="position:absolute; left:130px; margin-top:-36px">
<input type="text" value="Search web">
</form>
</div>
</div>



<div style="position:absolute; left:1098px; margin-top:-1500px">


<img src="indianflag.gif" alt="Indian Emblem" width:200px;
height:200px">


</div>

<div style="position:absolute; left:1098px; margin-top:-1000px">
<img src="emblem.jpg" alt="Indian Emblem" width:200px;
height:200px">


</div>
```

```html
<div style="position:absolute; left:1058px; margin-top:-500px">
<img src="55.jpg" alt="Donate" width:200px; height:200px">

</div>


<div style="position:absolute; left:60px; margin-top:-1150px">

<form action="submit">
<fieldset style="background:gray">
<legend style="background:white">Form No.987XAJ789</legend>
<h3 style="text-decoration:underline; font-size:33px; color:blue;
background:olive; height:60px"><a
href="https://"www.goi.com/applications/">Application form for
the post of AMD:</a></h3>

<fieldset style="width:950px; background:red">
<legend style="font-size:30px;
background:yellow">Instructions:</legend>
<textarea style="width:700px; height:100px">
Please read the instruction carefully before filling up the form.
In case of any doubt please Email:info@goi.com.

Last date for Online submission is:aa/dd/ffff. Any application
form submited after 6 PM aa/dd/ffff is considered void.

Application form can also be obtained directly from all head post
office. Last date to submit applciation form for people who are
living in remote places of India is on or before 6 PM aa/dd/ffff.

You can also down the application form from our website:
https://www.goi.com/applicationform/

</textarea><br> <br>
</fieldset>
<br> <br>


<label style="font-family:times new roman; font-size:20px">1)
Select Title</label>
<select>
<option>Mr</option>
<option>Miss</option>
<option>Mrs</option>
</select>
 <br> <br>
```

```html
<label style="font-family:times new roman; font-size:20px">2)
First Name:</label>
<input type="text" value"">
<label style="font-family:times new roman; font-size:20px">Middle
Initial:</label>

<input type="text" value"">

<label style="font-family:times new roman; font-size:20px">Last
Name:</label>

<input type="text" value""> <br> <br>

<label style="font-family:times new roman; font-size:20px">3)
Name of the Parent/Gaurdian:</label>

<input type="text" value"" style="Position:absolute;
left:300px"><br> <br>

<label style="font-family:times new roman; font-size:20px">4)
Address of the candidate for communication:</label><br> <br>

<label style="font-family:times new roman; font-
size:20px">Address (Line 1):</label>



<input type="text" value"" style="Position:absolute;
left:300px"><br> <br>

<label style="font-family:times new roman; font-
size:20px">Address (Line 2):</label>



<input type="text" value"" style="Position:absolute;
left:300px"><br> <br>

<label style="font-family:times new roman; font-
size:20px">City/Town/Village:</label>



<input type="text" value"" style="Position:absolute;
left:300px"><br> <br>
```

```html
<label style="font-family:times new roman; font-
size:20px">District:</label>


<input type="text" value="" style="Position:absolute;
left:300px"><br> <br>

<label style="font-family:times new roman; font-
size:20px">State:</label>


<input type="text" value="" style="Position:absolute;
left:300px"><br> <br>


<label style="font-family:times new roman; font-
size:20px">Pincode:</label>


<input type="text" value="" style="Position:absolute;
left:300px"><br> <br>

<label style="font-family:times new roman; font-size:20px">5)
Mobile Number:</label>


<input type="text" value="" style="Position:absolute;
left:300px"><br> <br>

<label style="font-family:times new roman; font-size:20px">6)
Email ID:</label>


<input type="text" value="" style="Position:absolute;
left:300px"><br> <br>


<label style="font-family:times new roman; font-size:20px">7)
Date of Birth:</label>
```

```
<Select style="Position:absolute; left:300px">
<optgroup label="Date">
<option>1</option>
<option>2</option>
<option>3</option>
<option>4</option>
<option>5</option>
<option>6</option>
<option>7</option>
<option>8</option>
<option>9</option>
<option>10</option>
<option>11</option>
<option>12</option>
<option>13</option>
<option>14</option>
<option>15</option>
<option>16</option>
<option>17</option>
<option>18</option>
<option>19</option>
<option>20</option>
<option>21</option>
<option>22</option>
<option>23</option>
<option>24</option>
<option>25</option>
<option>26</option>
<option>27</option>
<option>28</option>
<option>29</option>
<option>30</option>
<option>31</option>
</optgroup>

</select>


<Select style="Position:absolute; left:370px">

<optgroup label="Month">
<option>January</option>
<option>February</option>
<option>March</option>
<option>April</option>
<option>May</option>
<option>June</option>
<option>July</option>
```

```
<option>August</option>
<option>September</option>
<option>October</option>
<option>November</option>
<option>December</option>


</optgroup>
</select>


<Select style="Position:absolute; left:490px">

<optgroup label="Year">
<option>1976</option>
<option>1977</option>
<option>1978</option>
<option>1979</option>
<option>1980</option>
<option>1981</option>
<option>1982</option>
<option>1983</option>
<option>1984</option>
<option>1985</option>
<option>1986</option>
<option>1987</option>
<option>1988</option>
<option>1989</option>
<option>1990</option>
<option>1991</option>
<option>1992</option>
<option>1993</option>
<option>1994</option>
<option>1995</option>
<option>1996</option>
<option>1997</option>
<option>1998</option>
<option>1999</option>
<option>2000</option>
<option>2001</option>
<option>2002</option>
<option>2003</option>
<option>2004</option>
<option>2005</option>
<option>2006</option>
<option>2007</option>
<option>2008</option>
```

```html
<option>2009</option>
<option>2010</option>
<option>2011</option>
<option>2012</option>
<option>2013</option>
<option>2014</option>
</optgroup>
</select>


<br> <br>


<label style="font-family:times new roman; font-size:20px">8)
Gender:</label>

<select <Select style="Position:absolute; left:300px">>
<optgroup label="Sex">
<option>Male</option>
<option>Female</option>
<option>Cross Gender</option>

</optgroup>
</select>


<br> <br>

<label style="font-family:times new roman; font-size:20px">9)
Religion:</label>

<select style="Position:absolute; left:300px">
<optgroup label="Religion">
<option>Hindu</option>
<option>Christian</option>
<option>Muslim</option>
<option>others</option>



</optgroup>
</select>

<label style="font-family:times new roman; font-size:20px;
padding-left:390px">Others* </label>
```

```html
 <input type="text" value""
style="Position:absolute;left:600px"><br> <br>


<label style="font-family:times new roman; font-size:20px">10)
Nationality:</label>

<select style="Position:absolute; left:300px">
<optgroup label="Country">
<option>Indian</option>
<option>NRI</option>
<option>others</option>
</optgroup>
</select>

<label style="font-family:times new roman; font-size:20px;
padding-left:359px">Others* </label>
<input type="text" value"" style="Position:absolute;
left:600px"><br> <br>

<hr>
```

HTML & CSS - Curve

```html
<label style="font-family:times new roman; font-size:20px">By
clicking on the option "Submit" you agree to the terms and
conditions of Governement of India:</label><br><br>
<input type="button" value="submit" style="Position:absolute;
left:600px"> <input type="button" value="Decline"
style="Position:absolute; left:670px"><br> <br>

</fieldset>
</form>

</div>
</body>

</html>
```

Government Of India

My account, Login. Search web

Form No.987XAJ789

Application form for the post of AMD:

Instructions:

Please read the instruction carefully before filling up the form. In case of any doubt please Email:info@aoi.com.

Last date for Online submission is:aa/dd/ffff. Any application form submitted after 6 PM aa/dd/ffff is considered void.

Application form can also be obtained directly from all head post office. Last date

1) Select Title: Mr

2) First Name: _____ Middle Initial: _____ Last Name: _____

3) Name of the Parent/Gaurdian: _____

4) Address of the candidate for communication:

Address (Line 1): _____

Address (Line 2): _____

City/Town/Village: _____

District: _____

State: _____

Pincode: _____

5) Mobile Number: _____

6) Email ID: _____

7) Date of Birth: 1 January 1976

8) Gender: Male

9) Religion: Hindu Others* _____

10) Nationality: Indian Others* _____

By clicking on the option "Submit" you agree to the terms and conditions of Government of India:

submit decline

सत्यमेव जयते

SAMPLE 3: DESIGN: MULTIMEDIA – WEB SITE:

```html
<!doctype html>
<html>
<head>
<title>Design Multimedia Website</title>
<meta name="Design2" content="Sample web site design"/>
<style type="text/css">

body{
background-color:silver;
background-img:url('oxygen.jpg');

}

a:link{
color:Black;
}
a:hover{
color:red;
}
a:visited{
color:black;
}

#header{

font-size:21px;
font-family:Lucida Sans Typewriter;
font-weight:bold;
text-align:center;
border:1px solid olive;
background-image:url('foot.png');

}

#footer{

font-size:15px;
font-family:Lucida Sans Typewriter;
font-weight:bold;
margin-top:128px;
text-align:center;
border:1px solid olive;
background-image:url('foot.png');
```

```css
}

#mpicture{
padding-left:270px;
background:skyblue;
background-image:url('c.jpg');
height:400px;
position:absolute;
left:340px;
width:400px;
}

#audio{
position:absolute;
margin-top:2-00px;

}

#songlist{
border:1px solid orange;
width:314px;
height:400px;
padding-left:15px;
background-color:#ffcc99;
font-family:candara;
}

#list2{
border:1px solid orange;
width:278px;
height:340px;
position:absolute;
left:1011px;
margin-top:-403px;
background-color:skyblue;
font-family:candara;
padding:30px;


}


</style>
```

```
<body>

<div id=header>


<p><a href="https://www.xyz.com/music/">|Music|</a> <a
href="https://www.xyz.com/video/">|Video|</a> <a
href="https://www.xyz.com/movies/">|Movies|</a> <a
href="www.xyz.com/apps/">|APPS| </a></p>


<div style="position:absolute; left:1000px; margin-top:-39px;
font-size:12px">
<input type="text" value="Search web"style="position:absolute;
left:130px; margin-top:7px">


<p><a href="https://www.xyz.com/myaccount/">My Account,</a> <a
href="https://www.xyz.com/login/">Login,</a> </p>

</div>

</div>
```

HTML & CSS - Curve

```
<div id=mpicture>

</div>



<div id=songlist>
<h1 style="text-decoration:underline; font-size:17px">10 - Best
Albums:</h1><br>
<label>1) <a
href="https://www.xyz.com/dangerous/">Dangerous..</a></label><br>
<label>2) <a href="https://www.xyz.com/blackandwhite/">Black and
white..</a></label><br>
<label>3) <a
href="https://www.xyz.com/millinium/">Millenium..</a></label><br>
<label>4) <a href="https://www.xyz.com/babyonemoretime/">Baby one
more time..!</a></label><br>
<label>5) <a
href="https://www.xyz.com/blue/">Blue..</a></label><br>
<label>6) <a href="https://www.xyz.com/98degrees/">98 degrees and
rising..</a></label><br>
```

```html
<label>7) <a
href="https://www.xyz.com/westlife/">Westlife..</a></label><br>
<label>8) <a
href="https://www.xyz.com/cloudplay/">Cloudplay..</a></label><br>
<label>9) <a href="https://www.xyz.com/blackandblue/">Black &
Blue..</a></label><br>
<label>10) <a
href="https://www.xyz.com/creed/">Creed..</a></label><br>


</div>


<div id=list2>

<form>

<h2 style="text-decoration:underline"> Please leave your
feedback</h2>

<label>Title:</label>
<select>
<option>Mr</option>
<option>Miss</option>
<option>Mrs</option>
</select><br><br>

<label>Name:</label>
<input type="text" value=""
style="position:absolute;left:80px"><br><br>
<label>Email:</label>
<input type="text" value=""
style="position:absolute;left:80px"><br><br>
<label>Comments</label><br><br>
<textarea style="width:209px; height:60px">

</textarea> <br> <br>
<label> Click to submit:</label>
<input type="button" value="submit"
style="position:absolute;left:164px">
</form>

</div>
```

```html
<div id=audio>
<fieldset style="background:gray; width:920px; padding-
left:390px">
<legend style="FOnt-size:18px; background:silver; font-
family:copperplate">Play your song:
<progress>==</progress></legend>

<audio controls loop>

<source src="C:\Users\Santosh C J\Music\AUDIO\New\Yum.mp3">


</audio>

</fieldset>

</div>

<div id=footer>

<div style="font-size:13px">
<p><a href="https://www.xyz.com/home/">Home</a> | <a
href="https://www.xyz.com/aboutus/">About Us</a> | <a
href="https://www.xyz.com/conditionsofuse/">Conditions of Use</a>
|<a href="https://www.xyz.com/privacy/"> Privacy Policy</a> | <a
href="https://www.xyz.com/paymentpartnerts/">Payment Partners</a>
| <a href="https://www.xyz.com/disclaimer/">Disclaimer</a> | <a
href="https://www.xyz.com/sitemaps/">Site Maps</a> | <a
href="https://www.xyz.com/help/">Help/FAQs</a> | <a
href="https://www.xyz.com/contact/">Contact Us</a></p>
</div>

<div style="font-size:12px; font-family:times new roman">

<p>Copyright © 2014 XYZ Digital Media Entertainment Pvt. Ltd. All
Rights Reserved</p>

</div>
</div>



</body>

</html>
```

OUTPUT: SAMPLE 3: INTERNET EXPLORER:

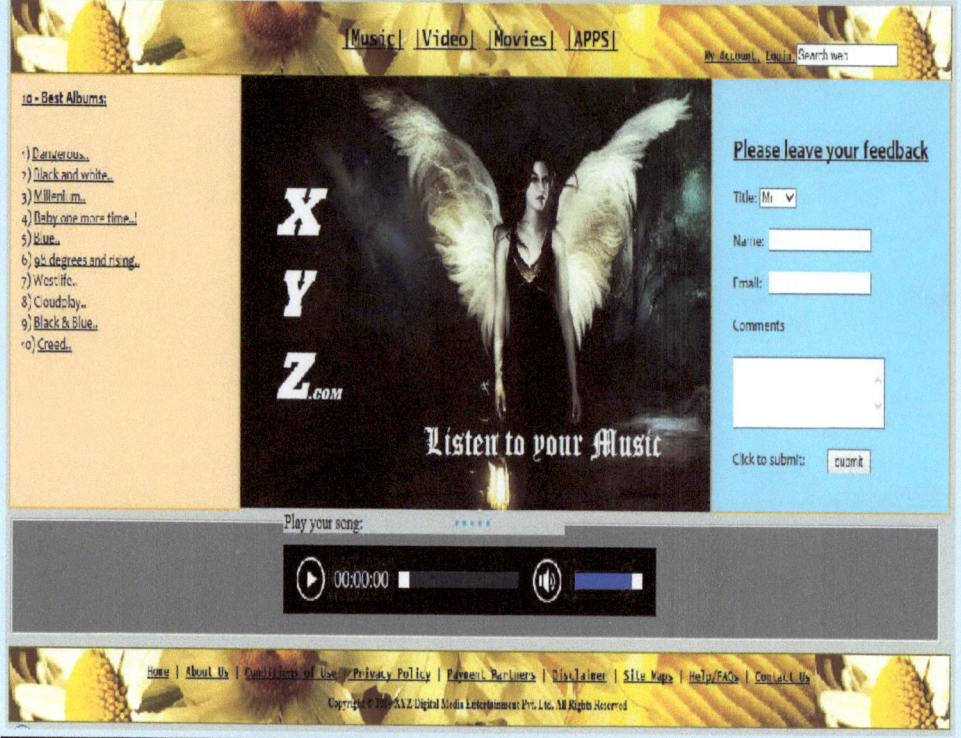

The winners in life think constantly in terms of I can, I will, and I am. Losers, on the other hand, concentrate their waking thoughts on what they should have or would have done, or what they can't do.

– Dennis Waitley

HTML & CSS - Curve

History

In 1991, a physicist named Berners-Lee released a document called as "HTML tags" which describes the basic and initial outline of HTML coding. HTML has evolved over a period of time and different versions are available. Each version has it unique character and can be used for different purpose.

HTML versions timeline

(Source: http://en.wikipedia.org/wiki/HTML)

November 24, 1995

- HTML 2.0 was published as IETF RFC 1866. Supplemental RFCs added capabilities:
- November 25, 1995: RFC 1867 (form-based file upload)
- May 1996: RFC 1942 (tables)
- August 1996: RFC 1980 (client-side image maps)
- January 1997: RFC 2070 (internationalization)

January 1997

HTML 3.2 was published as a W3C Recommendation. It was the first version developed and standardized exclusively by the W3C, as the IETF had closed its HTML Working Group in September 1996.

Initially code-named "Wilbur", HTML 3.2 dropped math formulas entirely, reconciled overlap among various proprietary extensions and adopted most of Netscape's visual markup tags. Netscape's blink element and Microsoft's marquee element were omitted due to a mutual agreement between the two companies. A markup for mathematical formulas similar to that in HTML was not standardized until 14 months later in MathML.

December 1997

HTML 4.0 was published as a W3C Recommendation. It offers three variations:

- Strict, in which deprecated elements are forbidden,
- Transitional, in which deprecated elements are allowed,
- Frameset, in which mostly only frame related elements are allowed ;

Initially code-named "Cougar", HTML 4.0 adopted many browser-specific element types and attributes, but at the same time sought to phase out Netscape's visual markup features by marking them as deprecated in favor of style sheets. HTML 4 is an SGML application conforming to ISO 8879 – SGML.

April 1998

HTML 4.0 was reissued with minor edits without incrementing the version number.

December 1999

HTML 4.01] was published as a W3C Recommendation. It offers the same three variations as HTML 4.0 and its last errata were published May 12, 2001.

May 2000

ISO/IEC 15445:2000] ("ISO HTML", based on HTML 4.01 Strict) was published as an ISO/IEC international standard. In the ISO this standard falls in the domain of the ISO/IEC JTC1/SC34 (ISO/IEC Joint Technical Committee 1, Subcommittee 34 – Document description and processing languages).

As of mid-2008, HTML 4.01 and ISO/IEC 15445:2000 are the most recent versions of HTML. Development of the parallel, XML-based language XHTML occupied the W3C's HTML Working Group through the early and mid-2000s.

HTML draft version timeline

October 1991

HTML Tags, an informal CERN document listing 18 HTML tags, was first mentioned in public.

June 1992

First informal draft of the HTML DTD, with seven subsequent revisions (July 15, August 6, August 18, November 17, November 19, November 20, November 22)

November 1992

HTML DTD 1.1 (the first with a version number, based on RCS revisions, which start with 1.1 rather than 1.0), an informal draft.

June 1993

Hypertext Markup Language was published by the IETF IIIR Working Group as an Internet-Draft (a rough proposal for a standard). It was replaced by a second version one month later, followed by six further drafts published by IETF itself that finally led to HTML 2.0 in RFC1866

November 1993

HTML+ was published by the IETF as an Internet-Draft and was a competing proposal to the Hypertext Markup Language draft. It expired in May 1994.

April 1995 (authored March 1995)

HTML 3.0 was proposed as a standard to the IETF, but the proposal expired five months later without further action. It included many of the capabilities that were in Raggett's HTML+ proposal, such as support for tables, text flow around figures and the display of complex mathematical formulas.

W3C began development of its own Arena browser as a test bed for HTML 3 and Cascading Style Sheets, but HTML 3.0 did not succeed for several reasons. The draft was considered very large at 150 pages and the pace of browser development, as well as the number of interested parties, had outstripped the resources of the IETF. Browser vendors, including Microsoft and Netscape at the time, chose to implement different subsets of HTML 3's draft features as well as to introduce their own extensions to it. (See Browser wars) These included extensions to control stylistic aspects of documents, contrary to the "belief [of the academic engineering community] that such things as text color, background texture, font size and font face were definitely outside the scope of a language when their only intent was to specify how a document would be organized." Dave Raggett, who has been a W3C Fellow for many years, has commented for example, "To a certain extent, Microsoft built its business on the Web by extending HTML features."

January 2008

HTML5 was published as a Working Draft (link) by the W3C.

Although its syntax closely resembles that of SGML, HTML5 has abandoned any attempt to be an SGML application and has explicitly defined its own "html" serialization, in addition to an alternative XML-based XHTML5 serialization.

May 2011

On 14 February 2011, the W3C extended the charter of its HTML Working Group with clear milestones for HTML5. In May 2011, the working group advanced HTML5 to "Last Call", an invitation to communities inside and outside W3C to confirm the technical soundness of the specification. The W3C is developing a comprehensive test suite to achieve broad interoperability for the full specification by 2014, which is now the target date for Recommendation.

- XHTML is a separate language that began as a reformulation of HTML 4.01 using XML 1.0. It is no longer being developed as a separate standard.
- XHTML 1.0, published January 26, 2000, as a W3C Recommendation, later revised and republished August 1, 2002. It offers the same three variations as HTML 4.0 and 4.01, reformulated in XML, with minor restrictions.
- XHTML 1.1, published May 31, 2001, as a W3C Recommendation. It is based on XHTML 1.0 Strict, but includes minor changes, can be customized, is reformulated using modules from Modularization of XHTML, which was published April 10, 2001, as a W3C Recommendation.
- XHTML 2.0 was a working draft, but work on it was abandoned in 2009 in favor of work on HTML5 and XHTML5. XHTML 2.0 was incompatible with XHTML 1.x and, therefore, would be more accurately characterized as an XHTML-inspired new language than an update to XHTML 1.x.
- XHTML syntax, known as "XHTML5.1", is being defined alongside HTML5 in the HTML5 draft.

HTML & CSS - Curve

- By Santosh C J.